SpringerBriefs in Applied Sciences and Technology

SpringerBriefs in Mathematical Methods

Mathematical Methods is a new series of SpringerBriefs devoted to non-standard and fresh mathematical approaches to problems in applied sciences. Compact volumes of 50 to 125 pages, each presenting a concise summary of a mathematical theory, and providing a novel application in natural sciences, humanities or other fields of mathematics. The series is intended for applied scientists and mathematicians searching for innovative mathematical methods to address problems arising in modern research. Examples of such topics include: algebraic topology applied in medical image processing, stochastic semigroups applied in genetics, or measure theory applied in differential equations.

Takashi Takiguchi

The Least Square Solution to an Overdetermined System

Theory and Applications

Springer

Takashi Takiguchi
Department of Mathematics
National Defense Academy of Japan
Yokosuka, Kanagawa, Japan

ISSN 2191-530X ISSN 2191-5318 (electronic)
SpringerBriefs in Applied Sciences and Technology
ISSN 2365-0826 ISSN 2365-0834 (electronic)
SpringerBriefs in Mathematical Methods
ISBN 978-981-92-0063-4 ISBN 978-981-92-0064-1 (eBook)
https://doi.org/10.1007/978-981-92-0064-1

Mathematics Subject Classification: Primary 15A29, 65F20, Secondary 65F22

This Springer imprint is published by the registered company Springer Nature Singapore Pte Ltd.
The registered company address is: 152 Beach Road, #21-01/04 Gateway East, Singapore 189721, Singapore

Dedicated to my mother, Masako, who has always been my support.

Preface

In this monograph, we study *how to give an approximate solution to an overdetermined system with no solution*, the meaning of which we shall explain using an example.

Consider the case that we measure the distances r_j between the unknown object (x, y, z) to be reconstructed by calculation and the measurement points (x_j, y_j, z_j), for $j = 1, 2, \cdots, m$, $(m \geq 3)$. If all measurements are precise, then we have a system

$$\begin{cases} (x - x_1)^2 + (y - y_1)^2 + (z - z_1)^2 = r_1^2, \\ \qquad \cdots \\ (x - x_m)^2 + (y - y_m)^2 + (z - z_m)^2 = r_m^2, \end{cases} \tag{1}$$

with a (in most cases, the unique) solution (x, y, z), however overdetermined $(m \gg 3)$ the system is. In practice, it is, however, impossible for each equation to be precise, since it is impossible to take measurements with no error. Therefore, in practice, what we almost always obtain by measurements is an overdetermined system with no solution if the system is overdetermined. Although the system obtained by measurements allows no solution, the object (x, y, z) really exists, which we would like to be approximately reconstructed. This is what is meant by "*how to give an approximate solution to an overdetermined system with no solution*."

It is more desirable in practice that more equations are included in the system (1), however overdetermined the system becomes, for which see Chap. 6.

In this monograph, we treat more generalized overdetermined systems than (1). We reduce the original overdetermined system with no solution to one consisting of $k \geq m \gg n$ linear equations:

$$\begin{cases} a_{11}x_1 + a_{12}x_2 + \cdots + a_{1n}x_n - s_1 = 0, \\ a_{21}x_1 + a_{22}x_2 + \cdots + a_{2n}x_n - s_2 = 0, \\ \quad \cdots\cdots\cdots \\ \quad \cdots\cdots\cdots \\ a_{k1}x_1 + a_{k2}x_2 + \cdots + a_{kn}x_n - s_k = 0, \end{cases} \tag{2}$$

We develop our theory in this monograph for such overdetermined systems. By our theory, the unique least-square solution to the system (2) turns out to be an approximate solution to the original system (1). The unique least-square solution to the reduced system (2) can be obtained by calculating the transpose ${}^tA$ of the coefficient matrix A of the reduced system (2), and the inverse $({}^tAA)^{-1}$ of the matrix ${}^tAA$, which is theoretically very easy. It is not so difficult to calculate $({}^tAA)^{-1}$ numerically by computers, if m is not so large, for example $m \leq 10$. However, if m and n are very large, for example, if we apply our theory to CT (computerized tomography), where $m \gg 10^8$ and $n \geq 10^7$, it is very costly and difficult to calculate $({}^tAA)^{-1}$ numerically by computers. We also discuss how to treat such large or huge systems, cheaply and easily.

The readers of this monograph are required to have learned elementary linear algebra, elementary calculus, and elementary probability theory. No other preliminary mathematical knowledge is required unless otherwise mentioned.

The author claims that there has been a rise in demand for this kind of monograph in this era of generative AI.

Tokyo
March 2026

Takashi Takiguchi

Acknowledgments The author expresses his gratefulness to Mr. Masayuki Nakamura, the editor of this book, who continuously helped and encouraged the author to write this book. The author is also grateful to the reviewers of this book for their valuable comments and suggestions.

Contents

Acronyms

2D	Two-dimensional
3D	Three-dimensional
ART	Algebraic reconstruction technique
CT	Computerized tomography
GLONASS	глобальная навигационная спутниковая система (Russian translation of "GNSS")
GNSS	Global navigation satellite system
GPS	Global positioning system
MRI	Magnetic resonance imaging
QZSS	Quasi-zenith satellite system
RC	Reinforced concrete
rebar	Reinforcing steel bar
USI	Ultrasonic imaging

Symbols

$E(X)$	The expectation of the random variable X
$\mathrm{Ker}(\varphi)$	The kernel of the mapping φ; If φ is a linear mapping from a linear space X to a linear space Y, then $\mathrm{Ker}(\varphi) = \{\boldsymbol{x} \in X \mid \varphi(\boldsymbol{x}) = \boldsymbol{0} \in Y\}$
$L^1(\mathbb{R}^2)$	The L^1 space on $\mathbb{R}^2$. $L^1(\mathbb{R}^2) = \{\varphi(x, y) \mid \int_{\mathbb{R}^2} \lvert\varphi(x, y)\rvert\, dx\, dy < \infty\}$
$L^2(\mathbb{R}^2)$	The L^2 space on $\mathbb{R}^2$. $L^2(\mathbb{R}^2) = \{\varphi(x, y) \mid \int_{\mathbb{R}^2} \lvert\varphi(x, y)\rvert^2\, dx\, dy < \infty\}$
$M_{mn}(\mathbb{R})$	The set of $m \times n$ matrices whose components are real numbers
$M_n(\mathbb{R})$	The set of $n \times n$ matrices whose components are real numbers
$\mathbb{N}$	The set of natural numbers $\mathbb{N} = \{1, 2, \cdots\}$
$\mathbb{R}$	The set of real numbers
$\mathbb{R}^n$	The n-dimensional Euclidean space. $\mathbb{R}^n = \{(x_1, x_2, \cdots, x_n) \mid x_j \in \mathbb{R},\ j = 1, 2, \cdots, n\}$
S^{n-1}	The unit sphere in $\mathbb{R}^n$. $S^{n-1} = \{(x_1, x_2, \cdots, x_n) \in \mathbb{R}^n \mid x_1^2 + x_2^2 + \cdots + x_n^2 = 1\}$
$\mathrm{supp} f$	The support of the function f; the smallest closed set containing $\{x \mid f(x) \neq 0\}$
${}^tM$	The transpose of the matrix M

Chapter 1
Introduction

Abstract The main purpose of this chapter is to introduce the motivation of this monograph. We introduce the general setting and how the topic is posed by practical requirements.

We first study the structure of solutions to the following overdetermined system allowing some solutions:

$$\begin{cases} F_1(x_1, x_2, \cdots, x_n) = 0, \\ \quad\cdots \\ F_m(x_1, x_2, \cdots, x_n) = 0, \end{cases} \tag{1.1}$$

where F_j are continuous (not necessarily algebraic) functions of n variable $\boldsymbol{x} = (x_1, x_2, \cdots, x_n)$ and $m > n$. The motivation of this study is based on the following system, frequently obtained by practical measurements:

$$\begin{cases} (x - x_1)^2 + (y - y_1)^2 + (z - z_1)^2 = r_1^2, \\ \quad\cdots \\ (x - x_m)^2 + (y - y_m)^2 + (z - z_m)^2 = r_m^2, \end{cases} \tag{1.2}$$

where $m > 3$. In the system (1.2), we measure the distances r_j between the unknown object (x, y, z) to be reconstructed and the measurement points (x_j, y_j, z_j). If all the measurement data are precise and they contain no errors, then the system must have a (in most cases, the unique) solution (x, y, z), however overdetermined the system is. In practice, it is, however, impossible for each equation to be precise, since it is impossible to take measurements with no errors. Therefore, in practice, what we obtain by measurements is the following system:

$$\begin{cases} (x - x_1)^2 + (y - y_1)^2 + (z - z_1)^2 = \rho_1^2, \\ \quad\cdots \\ (x - x_m)^2 + (y - y_m)^2 + (z - z_m)^2 = \rho_m^2, \end{cases} \tag{1.3}$$

T. Takiguchi, *The Least Square Solution to an Overdetermined System*, SpringerBriefs in Mathematical Methods, https://doi.org/10.1007/978-981-92-0064-1_1

where $\rho_1, \cdots, \rho_m$ ($\rho_j \approx r_j$) are real measured distances. Because of the errors $\epsilon_j = \rho_j^2 - r_j^2$ ($\rho_j = \sqrt{r_j^2 + \epsilon_j}$) included in the practical measurements, the possibility for the system (1.3) to admit a solution is zero if the system is overdetermined.

Even if the overdetermined system (1.3) admits no solution, the object (x, y, z) really exists. Therefore, we would like to approximately reconstruct the object (x, y, z) with the data ρ_j's containing errors and noise, which is the motivation to begin this research. For some practical applications of this research, see Chap. 3.

As we shall show in Chap. 6, for more accurate approximate reconstruction, it is more desirable to obtain more data, however overdetermined the system (1.2) would be.

In general, because of the errors in the measurements, we obtain the following system:

$$\begin{cases} F_1(x_1, x_2, \cdots, x_n) = \epsilon_1 \\ \cdots \\ F_m(x_1, x_2, \cdots, x_n) = \epsilon_m \end{cases} \tag{1.4}$$

instead of the system (1.1), for small unknown ϵ_j's.

In the next chapter, we study the structure of the solutions to (1.1), in order to study the structure of the least-square solutions to the system (1.4). The main purpose in the next section is to show that if the uniqueness of the least-square solution to the system (1.4) is proved then it must be a required approximate solution to the system (1.1) (Theorem 2.1 below). We also give a way to construct such an approximate solution to (1.1). It is more desirable in practice that it is possible to check whether the solution to (1.1) is unique only by the knowledge of the system (1.4) without knowing the system (1.1) itself, which is also studied in Chap. 2.

Remark 1.1 ("To Reconstruct the Solution" and "To Estimate the Solution") In order to mention the procedure to give the solution to an inverse problem, in mathematics, it is almost always mentioned *to reconstruct the solution*, while in engineering, it is often referred to *to estimate the solution*. It seems both meanings are not different. For the definition of inverse problems, see Definition 1.1 below. In this monograph, we mention the procedure to give the solution to an inverse problem *to reconstruct the solution*. In mathematics, the phrase *to estimate the solution* may be understood in another way "to evaluate the precision of the calculated solution applying some inequality". This inequality is called *an estimate of the solution*.

Definition 1.1 (Direct and Inverse Problems) A direct problem is a problem where we solve flow of time and/or cause-and-effect relations in a direct or a forward way. An inverse problem is a problem where we solve flow of time and/or cause-and-effect relations in an inverse or an backward way.

The following problems are direct problems:

- To solve a Cauchy problem of a differential equation, where we solve the future state of the time $t > 0$ from the initial state $t = 0$. Solution of this problem also directly solves the cause-and-effect relation.
- Weather forecasting, where we forecast the future weather using the information of the past and the present weather.

The following problems are inverse problems:

- To reconstruct the object point (x, y, z) in (1.2) by measuring the distances r_j's between the unknown object (x, y, z) to be reconstructed and the measurement points (x_j, y_j, z_j)'s. Solution of this problem inversely solves the cause-and-effect relation. In the framework of the system (1.2), the direct problem is to determine the r_j's by knowing (x, y, z) and (x_j, y_j, z_j)'s.
- When one has made a mistake, it is an inverse problem to think over what was a cause of the mistake, which is inversely thought over the cause-and-effect relation.

As we have seen in the above examples, the terms *direct and inverse problems* are widely applied to many problems, not restricted to mathematical ones.

Definition 1.2 In the system (1.2), if $m > n$ then the system is called *overdetermined*, if $m < n$ then the system is called *underdetermined*.

Exercises

Exercise 1.1 Choose the inverse problems from the followings:

(a) Solve the following Cauchy problem for the ordinary differential equation $\begin{cases} y'' + y = x \\ y(0) = y'(0) = 1 \end{cases}$

(b) If the unique solution to the ordinary differential equation $y'' + y = x$ is $y = \cos x + x$, then determine the initial values $y(0)$ and $y'(0)$.

(c) If the unique solution to the following Cauchy problem for the ordinary differential equation $\begin{cases} y'' + cy = x \\ y(0) = y'(0) = 1 \end{cases}$ is $y = \cos x + x$, then determine the constant coefficient c.

Exercise 1.2 Give some examples of (not necessarily mathematical) inverse problems.

Chapter 2
Structure and Approximation of the Solutions to Overdetermined Systems

Abstract In this chapter, we study the structure of the solutions to overdetermined systems with no solution, by whose application we also study how to approximately solve such overdetermined systems.

In this chapter, we study the structure of solutions to the system (1.1), which shall be applied to develop how to approximately solve such overdetermined systems without knowing the system (1.1) itself.

2.1 Structure of the Solutions to Overdetermined Systems

For the main purpose of this chapter, we deform the system (1.1) by the procedures called semi-equivalent deformations.

Definition 2.1 (Semi-equivalent Deformations) In the system (1.1), we call the following procedures *the semi-equivalent deformations*:

(i) Transpose some continuous terms in the left- or right-hand side to the other side in an equation.
(ii) Multiply both sides of one equation by the counterparts of another equation, or multiply both hand of an equation by a constant, where multiplication of a non-zero term and a zero term would not be allowed.
(iii) Add both sides of one equation to the counterparts of another.

Remark 2.1

- Semi-equivalent deformations are procedures to solve the system (1.1). They are different from the procedure to generate the ideal in $\mathbb{R}[x_1, \cdots, x_n]$ by $F_1, F_2, \cdots F_m$. For example, consider the system

T. Takiguchi, *The Least Square Solution to an Overdetermined System*, SpringerBriefs in Mathematical Methods, https://doi.org/10.1007/978-981-92-0064-1_2

$$\begin{cases} x^{30} + y^2 - z^3 = 0 \\ x^{15} - y - z = 0 \end{cases} \tag{2.1}$$

of the three variables x, y and z. We transpose $-y$ and $-z$ in the second equation to the right-hand side and take the square of both sides, from which we subtract the first equation to obtain

$$2y^2 + 2yz - z^3 + z^2 = 0. \tag{2.2}$$

We note that the polynomial $2y^2 + 2yz - z^3 + z^2$ would not belong to the ideal generated by $x^{30} + y^2 - z^3$ and $x^{15} - y - z$.

- There may exist several terms which become continuous as a total. For example, consider the Heaviside function

$$H(x) = \begin{cases} 0 & (x < 0), \\ \dfrac{1}{2} & (x = 0), \\ 1 & (x > 0). \end{cases} \tag{2.3}$$

The sum $H(x) + H(-x) \equiv 1$ is continuous as a total. In such a case, by (i) in Definition 2.1, we mean that the term $H(x)$ (or $H(-x)$) alone cannot be transposed to the other side. We have to treat the terms $H(x) + H(-x)$ as a sum (as a continuous function) in the semi-equivalent deformations.

Definition 2.2 (Semi-equivalent System) If a system named (D) of equations satisfies the following conditions:

(i) the system (D) is derived from the system (1.1) by semi-equivalent deformations;
(ii) the system (D) essentially contains the all equations in the system (1.1), by which we mean that any equation in the system (1.1) must be applied, at least once, in the procedure to derive the system (D) from the system (1.1) by semi-equivalent deformations,

then we call the system (D) *a semi-equivalent system* to the system (1.1).

Remark 2.2 Consider the following system:

$$\begin{cases} x + y = 3, \\ x + y = 1, \\ x \qquad = 2, \\ x \qquad = 0, \\ \qquad y = 2, \\ \qquad y = 0. \end{cases} \tag{2.4}$$

Subtract the last equation from the first and subtract the fourth equation from the first, then we have the following system:

$$\begin{cases} x = 3, \\ y = 3. \end{cases} \tag{2.5}$$

We note that the system (2.5) is not semi-equivalent to (2.4). It is because that the second, the third and the fifth equations in (2.4) have not been applied in the procedure to derive the system (2.5).

The condition (ii) in Definition 2.2 is imposed by practical requirement. As we have mentioned above, for more approximate reconstruction of the object (x, y, z), it is more desirable to include more equations in a semi-equivalent system, for which the condition (ii) in Definition 2.2 is imposed. The reason why it is more desirable to include as many equations as possible is discussed in Chap. 6.

Semi-equivalent deformations being procedures to solve the system (1.1), the following proposition is straightforward.

Proposition 2.1 *If $\boldsymbol{x} \in \mathbb{R}^n$ is a solution to the system (1.1) then it is also a solution to a semi-equivalent system to (1.1).*

By Proposition 2.1, the following proposition is obtained.

Proposition 2.2 *Assume that the system (1.1) admits a solution. If the solution to a semi-equivalent system (D) to the system (1.1) is unique then it must be the unique solution to (1.1). In this case, it is also proved that the system (1.1) has the unique solution.*

Proof We take a semi-equivalent system (D) to the system (1.1). By Proposition 2.1, we have the following inclusion relation:

$$\{\text{Solutions to (1.1)}\} \subset \{\text{Solutions to (D)}\} \tag{2.6}$$

Since the system (1.1) admits a solution by the hypothesis, if the solution to the system (D) is unique, then it must be also the unique solution to (1.1). □

Remark 2.3 Consider the following system (2.7):

$$\begin{cases} (x-1)^2 + (y-1)^2 = \rho^2, \\ (x+1)^2 + (y-1)^2 = \rho^2, \\ (x+1)^2 + (y+1)^2 = \rho^2, \\ (x-1)^2 + (y+1)^2 = \rho^2. \end{cases} \tag{2.7}$$

The system (2.7) allows no solution for $\rho \neq \sqrt{2}$, however, its semi-equivalent system

$$\begin{cases} x + y = 0, \\ x - y = 0, \\ x \quad\;\; = 0, \\ \quad\;\; y = 0, \end{cases} \tag{2.8}$$

admits the unique solution $x = y = 0$ for any ρ. Therefore, the strict inclusion of (2.6), that is,

$$\{\text{Solutions to (1.1)}\} \subsetneq \{\text{Solutions to (D)}\} \tag{2.9}$$

can happen. This is why we call the system defined in Definition 2.2 *a semi-equivalent system*, not *an equivalent system*, as well as calling the procedures defined in Definition 2.1 *the semi-equivalent deformations*, not *the equivalent deformations*. It is also interesting that the unique solution to the system (2.8) is the least-square solution to the system (2.7). It is an open problem whether this fact is generalized to the general cases, see Problem 2.1 below.

Problem 2.1 (An Open Problem About the Least-Square Solution to (1.1)) Assume that the system (1.1) allows no solution and its semi-equivalent system allows the unique solution. Under this assumption, study whether the unique solution to the semi-equivalent system is the unique least-square solution to the system (1.1).

It is interesting to study Problem 2.1, which may yield a new development of the research in semi-equivalent systems. Although the study of Problem 2.1 is interesting, the following assumption shall be assumed throughout of this monograph, which is satisfied in many practical applications.

Assumption 2.1 The system (1.1) admits a solution.

We claim that this assumption is very natural from the viewpoint of practical applications. For example, consider the system (1.2), which is a mathematical model to reconstruct the point (x, y, z) by the measurements of the distances between the point (x, y, z) and the various measurement points (x_j, y_j, z_j), $j = 1, \cdots, m$. The solution (x, y, z) must exist if all the measurements are exact, which is what Assumption 2.1 claims. Of course, it is not sure whether it is unique.

2.2 The Least Square Solutions to Overdetermined Systems of Linear Equations

In the rest of this chapter, we mainly treat linear systems. For readers' convenience, in Appendix below, we review elementary linear algebra, especially on matrix theory and on the analysis of underdetermined linear systems.

It often occurs in practice that the mathematical model of the problem (1.1) itself is an overdetermined system of linear equations, or that a semi-equivalent system to the system (1.1) is an overdetermined system of linear equations. Examples of the former are G. N. Hounsfield's overdetermined system of linear equations in the practicalization of the computerized tomography and the three-dimensional image reconstruction of asteroids for which we refer the readers to Sects. 3.1, 3.2 and [20, 22]. As an example of the latter, we take the system (1.2), where if we subtract the j-th equation from i-th one then we obtain an overdetermined system of $\frac{m(m-1)}{2}$ linear equations, for whose practical applications, see Sects. 3.3 and 3.4.

In this section, we focus on the case where an overdetermined system of linear equations is derived as a semi-equivalent system to the system (1.1).

Assume that a system of $k > m \gg n$ linear equations

$$\begin{cases} a_{11}x_1 + a_{12}x_2 + \cdots + a_{1n}x_n - s_1 = 0, \\ a_{21}x_1 + a_{22}x_2 + \cdots + a_{2n}x_n - s_2 = 0, \\ \quad\cdots\cdots\cdots \\ a_{k1}x_1 + a_{k2}x_2 + \cdots + a_{kn}x_n - s_k = 0, \end{cases} \tag{2.10}$$

or equivalently

$$Ax = s \tag{2.11}$$

is derived as a semi-equivalent system to (1.1). If the original system is (1.2), then by subtracting both hands sides of the i-th and j-th equations, we obtain the derived semi-equivalent system consisting of $k = \frac{m(m-1)}{2}$ linear equations.

In practice, the system (1.1), itself, would not be obtained. By errors and noise, what we can obtain is the system (1.4) with small error and noise ε_i in the i-th each equation. Consequently, we obtain the following overdetermined system (2.12) of linear equations:

$$\begin{cases} a_{11}x_1 + a_{12}x_2 + \cdots + a_{1n}x_n - s_1 = \varepsilon_1, \\ a_{21}x_1 + a_{22}x_2 + \cdots + a_{2n}x_n - s_2 = \varepsilon_2, \\ \quad\cdots\cdots\cdots \\ a_{k1}x_1 + a_{k2}x_2 + \cdots + a_{kn}x_n - s_k = \varepsilon_k, \end{cases} \tag{2.12}$$

or equivalently

$$Ax = \widetilde{s}, \tag{2.13}$$

where $\widetilde{s}_i = s_i + \varepsilon_i$ is the i-th component of $\widetilde{s}$ with small $|\varepsilon_i| \ll 1$ for $1 \le i \le k$.

It is often the case that a semi-equivalent system (2.10) to (1.1) has the unique solution since it contains no error, henceforth the original system (1.1) must also have the unique solution by Proposition 2.1. Because of the errors, however, the system (1.4) with errors $\epsilon_1, \cdots, \epsilon_n$ almost always becomes to allow no solution, neither does the semi-equivalent system (2.12) to the system (1.4).

In this monograph, we assume the followings.

Assumption 2.2 We assume that it is possible to derive an overdetermined linear system (2.10) (resp. (2.12)) as a semi-equivalent system to the system (1.1) (resp. (1.4)).

In this monograph, it is our main purpose to study the system (1.4) by application of the properties of the system (2.12). Therefore, Assumption 2.2 is the main assumption.

Assumption 2.3 The rank of the coefficient matrix A in (2.10) is n.

In practice, in order for Assumptions 2.2 and 2.3 to hold, it is essential to obtain sufficient measurement data for the system (1.1) to be overdetermined. For example, it is sufficient to obtain more than three measurements ($m \geq 4$) in the system (1.2) for Assumptions 2.2 and 2.3 to hold, since the unknown variables (x, y, z) to be reconstructed are three. In the example of the system (1.2), note also that it is sufficient to take $m = 4$ only for Assumption 2.3 to hold. Roughly speaking, in practice, if Assumption 2.2 is satisfied then Assumption 2.3 is also satisfied.

We also note that Assumptions 2.2 and 2.3 are satisfied in many practical applications by taking suitable and sufficient measurements, for which see the examples in Chap. 3 below. In this chapter, we study the structure of the least-square solutions to the overdetermined systems satisfying Assumptions 2.2 and 2.3. By Proposition 2.2 and Assumptions 2.2 and 2.3, we have the following proposition.

Proposition 2.3 *The solution to the overdetermined system (1.1) is unique.*

Remark 2.4 We note that it is possible to know whether the solution to the overdetermined system (1.1) is unique, only by the knowledge of the system (1.4) obtained by measurements, that is, we can check the unique solvability of the system (1.1), only knowing the system (1.4) obtained by measurements, without knowing the system (1.1)itself. It is because the unique solvability is determined by the rank of the coefficient matrix A of the system (2.10) (equivalently (2.11)), which is the same as the counterpart of the system (2.12) (equivalently (2.13)).

In order to construct an approximate solution to (1.4), which also serves as an approximate solution to (1.1), we study the overdetermined system (2.12). The system (2.12) allows a solution $\boldsymbol{x}$ if and only if

$$A\boldsymbol{x} - \widetilde{\boldsymbol{s}} = \boldsymbol{0}. \tag{2.14}$$

In view of (2.14), the least-square solutions are defined.

Definition 2.3 A vector $\boldsymbol{x} \in \mathbb{R}^n$ is called a least-square solution to (2.12) (or (2.13)) if and only if it minimizes the norm

$$\|A\boldsymbol{x} - \widetilde{\boldsymbol{s}}\|_{\mathbb{R}^m}, \tag{2.15}$$

where the norm $\|\boldsymbol{\alpha}\|_{\mathbb{R}^m} = \sqrt{\alpha_1^2 + \cdots + \alpha_m^2}$ is defined for $\boldsymbol{\alpha} = (\alpha_1, \cdots, \alpha_m) \in \mathbb{R}^m$.

We review the basic theory of the least-square solutions to a system of linear equations. Let us study the following system of linear equations.

Problem 2.2 Solve the following system of linear equations in $y_1, y_2, \cdots, y_n$:

$$\begin{cases} b_{11}y_1 + b_{12}y_2 + \cdots + b_{1n}y_n &= t_1, \\ b_{21}y_1 + b_{22}y_2 + \cdots + b_{2n}y_n &= t_2, \\ \cdots\cdots\cdots \\ b_{m1}y_1 + b_{m2}b_2 + \cdots + b_{mn}y_n = t_m, \end{cases} \tag{2.16}$$

or equivalently

$$B\boldsymbol{y} = \boldsymbol{t}, \tag{2.17}$$

where B is the coefficient matrix in (2.16), $\boldsymbol{y} = {}^t(y_1, y_2, \cdots, y_n)$ and $\boldsymbol{t} = {}^t(t_1, t_2, \cdots, t_m)$. For a matrix M (including a vector or $l \times 1$ ($l \in \mathbb{N}$) matrix), we denote its transpose by ${}^tM$.

For the least-square solutions to the system (2.16), the following propositions are known to hold.

Proposition 2.4 *Let $M_{mn}(\mathbb{R})$ be the set of $m \times n$ matrices whose components are real numbers. For $B \in M_{mn}(\mathbb{R})$, the following conditions are equivalent:*

(i) $\boldsymbol{y} \in \mathbb{R}^n$ *is a least-square solution to (2.16).*
(ii) *For any $\boldsymbol{z} \in \mathbb{R}^n$,*

$$\langle B\boldsymbol{z}, B\boldsymbol{y} - \boldsymbol{t}\rangle_{\mathbb{R}^m} = 0 \tag{2.18}$$

where $\langle \cdot, \cdot \rangle_{\mathbb{R}^m}$ represents the inner product in $\mathbb{R}^m$.
(iii) *The following equation holds:*

$${}^tBB\boldsymbol{y} - {}^tB\boldsymbol{t} = \boldsymbol{0}. \tag{2.19}$$

Proof Let us briefly review the proof of this proposition.

We first prove that (i) $\Leftrightarrow$ (ii): The condition (i) is equivalent to the following condition: For any $\mathbb{R}^n \ni \boldsymbol{z} \notin \mathrm{Ker}(B)$, it holds that the function

$$\varphi(t) := \|B(\boldsymbol{y} + t\boldsymbol{z}) - \boldsymbol{s}\|^2 = \langle B(\boldsymbol{y} + t\boldsymbol{z}) - \boldsymbol{s}, B(\boldsymbol{y} + t\boldsymbol{z}) - \boldsymbol{s}\rangle \tag{2.20}$$

takes its minimal value at $t = 0$. This condition is equivalent to $\varphi'(0) = 0$, since the function φ is a quadratic polynomial in t whose coefficients for the quadratic term is positive (see (2.21) below). The following representation,

$$\begin{aligned}\varphi(t) &= t^2\|Bz\|^2 + 2t\langle Bz, By - s\rangle_{\mathbb{R}^m} + \|By - s\|^2,\\ \varphi'(t) &= 2t\|Bz\|^2 + 2\langle Bz, By - s\rangle_{\mathbb{R}^m},\\ \varphi'(0) &= 2\langle Bz, By - s\rangle_{\mathbb{R}^m},\end{aligned} \tag{2.21}$$

yields that the condition $\varphi'(0) = 0$ is equivalent to the condition (ii), where we note that the condition (ii) trivially holds for any $z \in \mathrm{Ker}(B)$.

We then prove that (ii) ⇔ (iii): The condition (ii) is equivalent to the condition that for any $z \in \mathbb{R}^n$,

$$\langle Bz, By - s\rangle_{\mathbb{R}^m} = \langle z, {}^tB(By - s)\rangle_{\mathbb{R}^n} = 0, \tag{2.22}$$

which is equivalent to the condition $\mathbb{R}^n \ni {}^tB(By - s) = \mathbf{0}$. □

This proposition holds for any linear system (2.16). It does not matter whether the system is overdetermined or not.

Remark 2.5 In the proof of (i) ⇔ (ii) in Proposition 2.4, the function φ in (2.20) to minimize being quadratic with the positive coefficient of the quadratic term, we can represent its minimal state by completing the square, without applying differential calculus; for $\mathbb{R}^n \ni z \notin \mathrm{Ker}(B)$,

$$\begin{aligned}\varphi(t) &= t^2\|Bz\|^2 + 2t\langle Bz, By - s\rangle_{\mathbb{R}^m} + \|By - s\|^2\\ &= \|Bz\|^2\left(t + \frac{\langle Bz, By - s\rangle_{\mathbb{R}^m}}{\|Bz\|^2}\right)^2 - \frac{\langle Bz, By - s\rangle^2_{\mathbb{R}^m}}{\|Bz\|^2} + \|By - s\|^2.\end{aligned} \tag{2.23}$$

By (2.23), the function φ becomes minimal at $t = -\frac{\langle Bz, By - s\rangle_{\mathbb{R}^m}}{\|Bz\|^2}$. On the other hand, by definition, φ takes its minimal value at $t = 0$. Therefore,

$$\langle Bz, By - s\rangle_{\mathbb{R}^m} = 0 \tag{2.24}$$

which proves (i) ⇔ (ii) in Proposition 2.4.

In general, the following proposition holds for any matrix B.

Proposition 2.5

$$rank(B) = rank({}^tBB). \tag{2.25}$$

This proposition is proved by showing

$$\mathrm{Ker}(B) = \mathrm{Ker}({}^tBB), \tag{2.26}$$

where $\mathrm{Ker}(B)$ is the kernel of the matrix B.

By Proposition 2.5, the condition $\mathrm{rank}(B) = n$ is equivalent to the one that the matrix ${}^tBB$ is regular. Hence, the following proposition is clear from the condition (iii) in Proposition 2.4.

Proposition 2.6

(i) *For $B \in M_{mn}(\mathbb{R})$, there exists at least one least-square solution to (2.16) for any $s \in \mathbb{R}^m$.*
(ii) *If $\mathrm{rank}(B) = n$ for $B \in M_{mn}(\mathbb{R})$, then the least-square solution to the system (2.16) is uniquely given by the unique solution to the following system:*

$$ {}^tBBy = {}^tBt. \tag{2.27} $$

Let us go back to our problem to give the least-square solution to (1.4). By Assumptions 2.2 and 2.3, we have the following proposition.

Proposition 2.7 *The least-square solution to the overdetermined system (1.4) is uniquely determined.*

Therefore, we obtain the following theorem.

Theorem 2.1 *If the rank of the coefficient matrix in the system (2.10) (or (2.11)), equivalently in the system (2.12) (or (2.13)), is the same as the number of the unknown variables n, then the unique least-square solution to the system (1.4) is obtained by giving the unique solution to the system (2.12) (or (2.13)),*

$$ {}^tAAx = {}^tA\tilde{s}. \tag{2.28} $$

In this case, the unique solution to the system (2.28) is an approximation of the unique solution to the system (1.1).

Proof Since the first statement has been proved in the previous propositions, what we have to prove is the last one.

Since the rank of the coefficient matrix in the system (2.10) (or (2.11)), equivalently in the system (2.12) (or (2.13)), is the same as the number of the unknown variables n, by Assumption 2.1 and Proposition 2.2, the solution to the system (1.1) is unique. The facts that F_j's in (1.1) and (1.4) are continuous and that ϵ_j's in (1.4) are small imply that an approximate solution to the system (1.4) and the unique solution to the system (1.1) are close.

Since the least-square solution to the system (2.12) is an approximate solution to the system (1.4), the theorem is proved. □

Remark 2.6

(i) In Theorem 2.1, it is possible to know whether the solution to (1.1) is unique without knowing the system (1.1) itself. The knowledge of the system (1.4) with errors is sufficient. This fact is very important in practical applications, where we only obtain the system (1.4) with errors by measurements.

(ii) Even if the system (1.4) allows no solution, it can happen that its semi-equivalent system would allow the unique solution. Take the system (2.7) in Remark 2.3 for its example. The system (2.7) for $\rho \neq \sqrt{2}$ allows no solution, however, its semi-equivalent system (2.8) has the unique solution $x = y = 0$ for any ρ, as remarked in Remark 2.3. In such a case, the unique solution to the semi-equivalent system is an approximate solution to the system (1.1) by Theorem 2.1. We also note that this approximation is given in L^2–space.

(iii) When the condition number of the matrix ${}^tAA$ is large, we have to carefully calculate the least-square solution (2.29) below, for example, by application of high-precision numerical calculus.

(iv) Theorem 2.1 holds without Definition 2.2 (ii); however, we would like the least-square solution to the semi-equivalent system of linear equations with errors as close as possible to the unique solution to the original system with no error. It is known that if the error ε_j, $j = 1, \cdots, m$ in each equation in the system (2.12) of linear equations normally distributes, then the least-square solution to the system (2.12) almost surely converges to the real unique solution to the system (2.10) of linear equations with no error as $m \to \infty$, where m is the number of the linear equations included in the semi-equivalent system of linear equations (Theorem 4.2 in [22]). From the viewpoint of this fact, we would like to include as many equations in the original system as possible into the semi-equivalent systems. Therefore, Definition 2.2 (ii) is included in the definition of the semi-equivalent systems.

(v) We claim the simplicity of our idea. As a method to construct a least-square solution to a system of linear equations, the idea of the pseudo-inverse matrix or the Moore–Penrose inverse matrix [12, 15] is well known. By virtue of Assumption 2.3, we have given the Moore–Penrose inverse matrix in the solution of (2.28) without introducing its definition. The solution to (2.28) is given by

$$\boldsymbol{x} = ({}^tAA)^{-1}\,{}^tA\tilde{\boldsymbol{s}}. \tag{2.29}$$

It is known that $({}^tAA)^{-1}\,{}^tA$ is the Moore–Penrose inverse if $\operatorname{rank}(A) = n$, for which see the next section. Therefore, under the natural condition, Assumption 2.3, satisfied in many practical applications, the idea of the Moore–Penrose inverse can be introduced without introducing additional knowledge of mathematics. It is very important for many non-mathematical researchers engaging in practical mathematical applications, to introduce useful mathematical ideas with introducing the least necessary mathematical knowledge. In order to apply our main theorem in practice, the non-mathematical researchers need not know the idea of the Moore–Penrose inverse matrix introduced in the next section; it is sufficient to understand Theorem 2.1.

See [5, 20, 22] for more detail about the least-square solutions to the overdetermined systems of linear equations. If we can derive the system (2.12) of linear equations from the system (1.4) by semi-equivalent deformations, then we have simple methods to analyze and approximately solve it, which are very convenient and helpful.

For the proof of Theorem 2.1, we have assumed Assumption 2.3. In many practical applications, it is possible to obtain $\mathrm{rank}(A) = n$ by taking suitable and sufficiently many measurements, for which, see the examples in the next chapter. In such cases, the least-square solution to the system (1.4) given by (2.28) is a good approximate solution for the object $(x_1, \cdots, x_n)$ in (1.1) to be reconstructed.

2.3 The Moore–Penrose Inverse Matrix

In the previous section, we have mentioned the idea of the Moore–Penrose inverse matrix. Although it is not necessary to understand the idea of the Moore–Penrose inverse matrix for the practical applications of our theory, we shall shortly introduce it. The readers interested in practical applications, not in mathematical theory, can skip this section.

Definition 2.4 (The Moore–Penrose Inverse Matrix) For $A \in M_{mn}(\mathbb{R})$, its *Moore–Penrose inverse matrix* $A^\dagger \in M_{nm}(\mathbb{R})$ is defined as the matrix satisfying the following conditions:

(α) $AA^\dagger A = A$,
(β) $A^\dagger AA^\dagger = A^\dagger$,
(γ) ${}^t(A^\dagger A) = (A^\dagger A)$,
(δ) ${}^t(AA^\dagger) = (AA^\dagger)$.

For the Moore–Penrose inverse matrices, the following properties are known.

Proposition 2.8 *The Moore–Penrose inverse of a matrix $A \in M_{mn}(\mathbb{R})$ is unique if it exists.*

Proof Assume that for a matrix $A \in M_{mn}(\mathbb{R})$, there exist two matrices $A^\dagger, \widetilde{A} \in M_{nm}(\mathbb{R})$ satisfying the conditions (α)–(δ) in Definition 2.4. The idea for the proof is to show that

$$A^\dagger = A^\dagger A\widetilde{A} = \widetilde{A}. \tag{2.30}$$

In fact, for $A^\dagger$, it holds that

$$\begin{aligned} A^\dagger &= A^\dagger AA^\dagger = A^{\dagger\, t}(AA^\dagger) = A^{\dagger\, t}(A\widetilde{A}AA^\dagger) \\ &= A^{\dagger\, t}(AA^\dagger)\,{}^t(A\widetilde{A}) = A^\dagger AA^\dagger A\widetilde{A} = A^\dagger A\widetilde{A}. \end{aligned} \tag{2.31}$$

In (2.31), we have applied, (β) in Definition 2.4 for the first equation, (δ) for the second, (α) for the third, ${}^t(BC) = {}^tC\,{}^tB$ (for $B, C \in M_n(\mathbb{R}^n)$) for the fourth, ($\delta$) for the fifth, and ($\alpha$) for the last.

The representation $\widetilde{A} = A^{\dagger}A\widetilde{A}$ can be obtained in a similar way, where the condition (c) plays an important part instead of (d) in (2.31), for more in detail, see Exercise 2.3 below.

The fact $A^{\dagger} = \widetilde{A}$ proves that the Moore–Penrose inverse of a matrix $A \in M_{mn}(\mathbb{R})$ is unique if it exists. □

Proposition 2.9 *For $A \in M_{mn}(\mathbb{R})$ $(m \geq n)$ with rank$(A) = n$, its Moore–Penrose inverse $A^{\dagger}$ is given by*

$$A^{\dagger} = ({}^{t}AA)^{-1}\,{}^{t}A. \tag{2.32}$$

In fact, we can check the conditions (α)–(δ) in Definition 2.4, by directly substituting (2.32), for which see Exercise 2.4 below.

See [15] for more about the Moore–Penrose inverses of matrices.

Exercises

Exercise 2.1 For $B \in M_{mn}(\mathbb{R}),\ \boldsymbol{x} \in \mathbb{R}^{n},\ \boldsymbol{y} \in \mathbb{R}^{m}$, show

$$\langle B\boldsymbol{x}, \boldsymbol{y}\rangle_{\mathbb{R}^m} = \langle \boldsymbol{x}, {}^{t}B\boldsymbol{y}\rangle_{\mathbb{R}^n}. \tag{2.33}$$

Exercise 2.2 Prove Proposition 2.5.

Exercise 2.3 (Complement of the Proof of Proposition 2.8)

(a) Show ${}^{t}(BC) = {}^{t}C\,{}^{t}B$ for $B, C \in M_n(\mathbb{R}^n)$.
(b) Prove $\widetilde{A} = A^{\dagger}A\widetilde{A}$ in the proof of Proposition 2.8.
(c) Consider another proof of Proposition 2.8.

Exercise 2.4 Prove Proposition 2.9.

Chapter 3
Examples of Practical Applications

Abstract In this chapter, we introduce practical applications of the least-square solution to the overdetermined system (1.1).

In this chapter, we introduce practical applications of the theory established in the previous chapter. In practical applications, it often occurs that the system (1.1) itself is an overdetermined system of linear equations, or that a semi-equivalent system to the system (1.1) is an overdetermined system of linear equations, both of which shall be introduced in this chapter. We call the former type problems *type I problems* and the latter ones *type II problems*. For examples of the type I problems, see Sects. 3.1 and 3.2. For examples of the type II problems, see Sects. 3.3 and 3.4.

3.1 Practical Implementation of CT by G. N. Hounsfield

In this section, we introduce the idea by Godfrey Newbold Hounsfield for the practical implementation of computerized tomography (CT). He succeeded in practical implementation of the medical CT, for which he shared the 1979 Nobel Prize for Physiology or Medicine with Allan McLeod Cormack.

3.1.1 Review of Differential

In this subsection, we review the idea of *the differential*, which is a key idea to represent phenomena mathematically. We begin the review the definition of the differentiability and the total differentiability.

Definition 3.1 (Differentiability) A function f of one-variable is differentiable at the point x_0 if and only if the following limit

T. Takiguchi, *The Least Square Solution to an Overdetermined System*, SpringerBriefs in Mathematical Methods, https://doi.org/10.1007/978-981-92-0064-1_3

$$\lim_{h \to 0} \frac{f(x_0 + h) - f(x_0)}{h} = f'(x_0) \tag{3.1}$$

exists. $f'(x_0)$ is called the differential coefficient of f at x_0.

The idea of the differentiability is very familiar.

Definition 3.2 (Total Differentiability) A function f of one-variable is totally differentiable at the point x_0 if and only if

$$\begin{cases} f(x_0 + h) - f(x_0) = \alpha h + g(h) \\ \lim_{h \to 0} \dfrac{g(h)}{h} = 0 \end{cases} \tag{3.2}$$

holds for some constant α and some one-variable function $g(h)$.

The idea of the total differentiability is not so familiar to those who do not major in mathematics. Even to mathematical students, the idea of the total differentiability of the one-variable functions may not be so familiar. Mathematical students may know well the idea of the total differentiability of the several-variable functions.

The idea of the total differentiability of one-variable functions being not so familiar, it is important to introduce the idea of *differential* or *infinitesimal*, which is a key idea to represent phenomena mathematically.

In the rest of this subsection, let us study total differentiability and *the differential* or *the infinitesimal*, which will be applied to give the mathematical model of CT in the next subsection.

Proposition 3.1 *If a function f is totally differentiable at the point x_0, then it is differentiable at x_0 and the constant in α in (3.2) is the differential coefficient of f at x_0, that is,*

$$\alpha = f'(x_0). \tag{3.3}$$

This proposition is proved by dividing both sides of the first equation in (3.2) by h and tending $h \to 0$.

Theorem 3.1 *For a function f of one-variable, the following conditions are equivalent.*

(i) *The function f is differentiable at the point x_0.*
(ii) *The function f is totally differentiable at the point x_0.*

Remark 3.1 The idea of the total differentiability of one-variable functions is not frequently taught in the class of elementary calculus. It seems that it is because of Theorem 3.1, that is, for a one-variable function, the differentiability and the total differentiability are equivalent. On the other hand, for a several variable function, the partial the differentiability and the total differentiability are completely different.

Assume that the function $f(x)$ of one-variable is totally differentiable at the point x, equivalently, differentiable at the point x by Theorem 3.1. By the total differentiability and Proposition 3.1, we obtain the following representation

$$\begin{cases} f(x+h) - f(x) = f'(x)h + g(h) \\ \displaystyle\lim_{h \to 0} \frac{g(h)}{h} = 0 \end{cases} \tag{3.4}$$

for some one-variable function $g(h)$.

Definition 3.3 (Differential (Infinitesimal)) We define the differential dx of the variable x as the infinitesimal (infinitely small) change in the variable x, which satisfies

$$|dx| = 0 \quad \text{and} \quad dx \neq 0. \tag{3.5}$$

Note that the differential dx is a *concept* not a *number*. If dx is a number, then $|dx| = 0$ implies $dx = 0$.

In (3.4), let h to the infinitesimal dx. By the first line in (3.4),

$$f(x+dx) - f(x) = f'(x)\,dx + g(dx). \tag{3.6}$$

We denote the left-hand side in (3.6) by

$$df(x) = f(x+dx) - f(x) \tag{3.7}$$

and call it *the differential of the function f at the point x*. By the second line in (3.4), we also obtain

$$\frac{g(dx)}{dx} = 0 \quad \Rightarrow \quad g(dx) = 0 \cdot dx = 0. \tag{3.8}$$

In the right-hand equation in (3.8), though $dx \neq 0$, 0 is 0. Hence, their product gives 0.

By (3.6), (3.7), and (3.8), we obtain the following representation:

$$df(x) = f(x+dx) - f(x) = f'(x)\,dx. \tag{3.9}$$

In (3.9), the differential $df(x)$ of f at x is proportional to the differential dx of x, whose proportional constant is $f'(x)$. Therefore, $f'(x)$ is called *the differential coefficient of the function f at x*.

The representation (3.9) is called *the differential representation of the function f at the point x*. The differential representation (3.9) is essential for the mathematical presentation of phenomena.

Let us introduce an example to give a mathematical representation of a phenomenon. Let $t \geq 0$ be the time, $u(t)$ be the number of the bacteria (or the amount of the investment) and $r(t)$ be the rate of increase. Note that $r(t)$ can take negative values.

It is not easy to give a time-global mathematical representation of a phenomenon changing each and every second; however, if we try to give a mathematical representation in the infinitesimal time interval $[t, t + dt]$, then it is possible.

In the infinitesimal time interval $[t, t + dt]$, the increased number of the bacteria (or the increased amount of the investment) is represented by $u(t + dt) - u(t)$, which is proportional to $u(t)\,dt$ with its proportional constant $r(t)$. Note that in the infinitesimal time interval $[t, t + dt]$, whose length is $|dt| = 0$, we can take $r(t)$ for *a constant*.

Therefore, we obtain

$$u'(t)\,dt = u(t + dt) - u(t) = u(t)r(t)\,dt, \tag{3.10}$$

which yields

$$u'(t) = u(t)r(t). \tag{3.11}$$

Equation (3.11) is a typical example of an ordinary differential equation of separation of variables type.

3.1.2 Mathematical Model of CT

In this subsection, we review the mathematical model of CT and introduce some problems for CT to be practicalized. The term "CT" is the abbreviation for "computerized tomography" and the term "tomography" is a compound made of "$\tau o\mu o\sigma$ (=tomos, section) + graphy", hence, interpreted literally, by "computerized tomography", it is meant "to draw a picture of the section by a computer". It does not seem difficult to understand what it is about. By summing up, what we see when we are examined by CT are computer graphics of the sections of our bodies, not simple photographs. Since the principle of CT is very simple and understandable, with the introduction of which, let us begin. We first introduce the properties of the X-ray projected at the human body for CT.

(i) The X-ray rectilinearly propagates.
(ii) The X-ray is attenuated by passing through the human body. The amount of the attenuation is proportional to the density of the human body.
(iii) When the X-ray passes through a homogeneous object, the amount of its attenuation is proportional to the length of the segment in the object where the X-ray propagates.

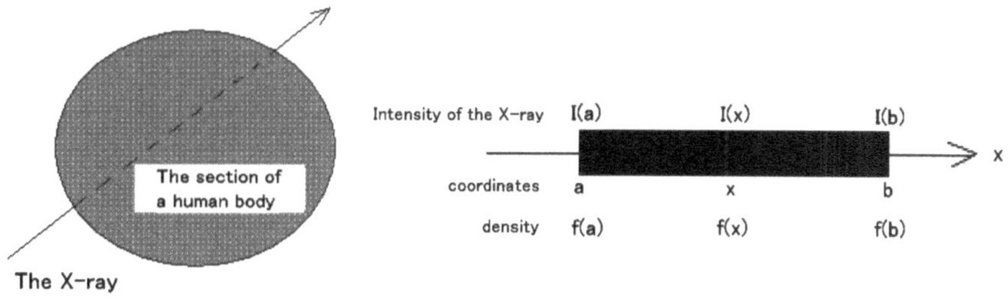

Fig. 3.1 Image of CT scan

Strictly speaking, the X-ray propagates in a cone whose solid angle is very small; we cannot say that the X-ray rectilinearly propagates. A small amount of the X-ray would be reflected and scattered when it is projected to the human body. In addition to the above, there are some other reasons why the above properties (i)–(iii) are not strictly right. The amount of the X-ray which does not follow the above properties (i)–(iii) is extremely small and its effect is much smaller than the measurement and numerical errors. Therefore, we can assume the above properties (i)–(iii) in view of practical application of CT. Let us consider the section of the human body in Fig. 3.1 and the X-ray projected at it. Since the X-ray rectilinearly propagates, we take out the line where the X-ray propagates and let it be the x-axis. We also let the point where the X-ray enters the human body as the point $x = a$ and the point where the X-ray goes out of the human body as the point $x = b$.

For $a \leq x \leq b$, let $f(x)$ be the density of the human body and $I(x)$ be the intensity of the X-ray. The differential $dI(x)$ of the intensity of the X-ray is given by

$$dI(x) = I(x + dx) - I(x) = I'(x)\,dx, \tag{3.12}$$

while by virtue of the above properties (i)–(iii), the X-ray propagating from the point $x \in (a, b)$ by the length dx is represented by

$$dI(x) = I(x + dx) - I(x) = I'(x)\,dx = -f(x)I(x)\,dx. \tag{3.13}$$

For simplicity, we have let the constant of proportionality in (iii) be 1. Note that the sign of $dI(x)$ is negative since the X-ray is attenuated; the amount is proportional to the density $f(x)$ of the human body, the intensity $I(x)$ of the X-ray and the length dx where the X-ray propagates. The simple representation (3.13) is possible since the length dx is infinitesimal.

By (3.13), we obtain

$$\log I(a) - \log I(b) = \int_a^b f(x)\,dx. \tag{3.14}$$

Therefore, by the measurements of $I(a)$ and $I(b)$, we obtain the integration of $f(x)$ in the interval $[a, b]$. In the practical CT scan, the problem is to reconstruct the density $f(x, y)$ of the human body by the knowledge of the intensity of the X-ray before and after it is projected at the human body along any line in the plane. Therefore, the problem of CT is formulated as follows.

Problem 3.1 Reconstruct the function $f(x, y)$ defined on $\mathbb{R}^2$ by its line integrals $\int_l f(x, y)\,dl$ along all lines l in $\mathbb{R}^2$.

In the above problem, it is sufficient to assume $f \in L^1 \cap L^2(\mathbb{R}^2)$ which is satisfied in the practical CT. If you have not learned Lebesgue integral, it is sufficient to assume that f is piecewise continuous. When the X-ray is projected at the human body, a phenomenon occurs whereby the X-ray is attenuated, whose cause is the density $f(x, y)$ of the human body. Therefore, Problem 3.1 is an inverse problem to reconstruct the cause $f(x, y)$ of the phenomenon by the knowledge of the result $\int_l f(x, y)\,dl$ of the phenomenon, by which we mean that this problem is an inverse problem to solve the inverse cause-and-effect relation.

In general, for a function $f(x)$ defined on $\mathbb{R}^n$ and $d < n$, its d-dimensional Radon transform $R_d f(\xi)$ is defined as its integration on the d-dimensional plane ξ. For $d = n-1$, we omit the term $n-1$ and write Rf, which is called the Radon transform of f. Problem 3.1 is a problem of the Radon transform on $\mathbb{R}^2$. For the general theory of the Radon transform, cf. [4, 7, 8]. As a reference for the Radon transform in connection with CT, we recommend [13]. The transform is named after J. Radon, who in 1917 [17] gave inversion formulas for the Radon transform on $\mathbb{R}^n$ for $n = 2, 3$. His original paper was written in German, but we can find its English translation in the appendix in [4]. It is a pity that the inversion formula proved by J. Radon did not contribute to the practical implementation of CT at the al stage. Both A. M. Cormack and G. N. Hounsfield, who won the Nobel Prize in Physiology or Medicine in 1979 for the practical implementation of CT, did not modify or apply the reconstruction formula by J. Radon. For the present-day CT, however, modification of Radon's inversion formula is applied to establish suitable reconstruction algorithms, which are known as *the filtered back projection techniques* (cf. [13] for example).

If the reconstruction formula by J. Radon had been widely known, would the practical implementation of CT have been much earlier? It is not a straightforward story, as we shall see. First, let us remark that in practical CT, the data obtained by the measurements are only approximate values of the line integral of $f(x, y)$, not the true line integrals themselves, because of the errors in the measurements, the approximation by finite data, numerical rounding errors by computers and so on. J. Radon's reconstruction formula being not stable under errors, some regularization and discretization treatments were necessary for the practical implementation of CT, one of which was realized by the filtered back projection techniques for today's medical CT.

Problem 3.2 Assume that for two functions $f_1(x, y)$, $f_2(x, y)$ defined on $\mathbb{R}^2$ and for any line $l \subset \mathbb{R}^2$, their Radon transforms are close, that is,

$$Rf_1(l) \approx Rf_2(l) \quad \text{for } \forall l \subset \mathbb{R}^2. \tag{3.15}$$

In this case, is it true that

$$f_1(x, y) \approx f_2(x, y) \quad \text{for } \forall(x, y) \in \mathbb{R}^2? \tag{3.16}$$

Before studying this problem, we have to study the following uniqueness problem.

Problem 3.3

$$Rf_1 \equiv Rf_2 \Rightarrow f_1 \equiv f_2? \tag{3.17}$$

This problem is positively solved in practical CT, for which see Exercise 3.3 below. In general, for the uniqueness of the Radon transform to hold, it is sufficient to assume that $f \in L^1(\mathbb{R}^2)$. Its proof is an exercise of the Fourier transform (cf. [21, 27], for example). An answer to Problem 3.2 depends on the classes the functions f_1 and f_2 belong to. It is known that, in the framework of the function spaces in accordance with the measurement data in practical CT, the answer to Problem 3.2 is negative, that is, the density distribution of the section of the human body reconstructed by the reconstruction formula by J. Radon with the observed data which must contain errors may be far from the original density distribution $f(x, y)$. It is known for the Radon transform that any set of finite data would not assure the uniqueness. It is, however, known that if we obtain sufficiently many data of the Radon transform then we can reconstruct a sufficiently approximate solution to the original function f, for which, confer [7].

Taking the above discussion into account, the problem of practical CT is formulated as follows.

Problem 3.4 Let $m \in \mathbb{N}$ be sufficiently large and $l_1, l_2, \cdots, l_m$ be mutually different lines in $\mathbb{R}^2$. Construct an approximate solution of the density distribution $f(x, y)$ from the observed data $F_1, F_2, \cdots, F_m$ $(F_i \approx Rf(l_i),\ i = 1, 2, \cdots, m)$ necessarily containing errors.

Today, there are many approaches to this problem, among which we shall introduce the first one by G. N. Hounsfield in the next subsection. We note again that the idea called *filtered back projection* mentioned above, which is other than the idea by G. N. Hounsfield, is applied to today's medical CT.

3.1.3 The Idea by G. N. Hounsfield

In numerical analysis, we usually discretize a mathematical formula obtained by mathematical analysis for the practically applicable implementation; however, the essence of the idea by G. N. Hounsfield is to discretize the model itself. It is very interesting and worked very well at the initial stage of the practical implementation of CT. We can assume that supp f is compact, or a bounded closed set, and $f \in L^1(\mathbb{R}^2)$ since f is a density distribution of the human body. Cover supp f with n squares $c_1, c_2, \cdots, c_n$, whose sides are of length 1 and parallel to the x- or y-axis and any pair of whose interiors are mutually disjoint. These squares are called *pixels*. We approximate this function f by a function $g(x, y)$ defined as

$$g(x, y) = \sum_{j=1}^{n} x_j \chi_{c_j}(x, y), \tag{3.18}$$

where $\chi_{c_j}(x, y)$ is the characteristic function of the square c_j and x_j are unknowns. For example, if we take x_j as the integral mean in the square c_j

$$x_j = \frac{1}{|c_j|} \int_{c_j} f(x, y)\, dx\, dy, \tag{3.19}$$

then it seems easy to understand G. N. Hounsfield's idea, where $|c_j|$ is the area of the square c_j. Let us call the function $g(x, y)$ *a pixel function*. G. N. Hounsfield tried to reconstruct an approximation of the pixel function $g(x, y)$, neither $g(x, y)$ itself nor $f(x, y)$ itself, from the observed data necessarily containing errors. This idea is very nice. Since the best we can hope is to obtain an approximation of f, not to reconstruct f itself, it is very flexible to approximate f by a suitably simple function g so that the problem would be simplified. Assume that m X-rays, $I_1, I_2, \cdots, I_m$, are projected at the human body. We assume that for $i = 1, 2, \cdots, m$, the intensity of I_i before the projection is I_i^0 and after the projection is I_i^1. By $l_1, l_2, \cdots, l_m$, we denote the lines where the X-ray $I_1, I_2, \cdots, I_m$ rectilinearly propagates, respectively and by a_{ij}, $i = 1, 2, \cdots, m$, $j = 1, 2, \cdots, n$, we denote the length of $l_i \cap c_j$. Let $s_i := \log I_i^0 - \log I_i^1$, $i = 1, 2, \cdots, m$. By errors in approximation of the density function $f(x, y)$ by a pixel function, error in measurements, numerical rounding errors by computers and so on, what we can obtain by measurements are

$$\widetilde{s_i} = s_i + \varepsilon_i \tag{3.20}$$

with small errors $|\varepsilon_i| \ll 1$ mentioned above for $i = 1, 2, \cdots, m$, not s_i's themselves. Therefore, the problem we have to solve turns out to be the following one.

Problem 3.5 Solve or give an approximate solution to the following system of linear equations in $x_1, x_2, \cdots, x_n$:

$$\begin{cases} a_{11}x_1 + a_{12}x_2 + \cdots + a_{1n}x_n &= \widetilde{s}_1, \\ a_{21}x_1 + a_{22}x_2 + \cdots + a_{2n}x_n &= \widetilde{s}_2, \\ \cdots\cdots\cdots \\ a_{m1}x_1 + a_{m2}x_2 + \cdots + a_{mn}x_n &= \widetilde{s}_m, \end{cases} \tag{3.21}$$

or equivalently

$$Ax = \widetilde{s}, \tag{3.22}$$

where $\widetilde{s} = {}^t(\widetilde{s}_1, \widetilde{s}_2, \cdots, \widetilde{s}_m)$.

The system (3.21) (equivalently, (3.22)) is the same as the system (2.12) (equivalently, (2.13)).

In the medical CT, we have to treat a huge system like $m \geq 10^8,\ n \geq 10^6$. The fact $m \gg n$ and the effect of the errors we mentioned above, the system (3.21) must have no solution. Although the system (3.21) has no solution, the density distribution $f(x, y)$ of the human body exists with no doubt, which must be approximately solved.

By the least-square solution

$$x^l = {}^t(x_1^l, x_2^l, \cdots, x_n^l) \tag{3.23}$$

to the system (3.21) (or (3.22)), we obtain an approximation pixel function

$$g_l(x, y) = \sum_{j=1}^{n} x_j^l \chi_{c_j}(x, y), \tag{3.24}$$

of the original density distribution function $f(x, y)$. For this approximation pixel function $g_l(x, y)$, the following theorem holds:

Theorem 3.2 *The pixel function (3.24) is an approximate solution to the original density distribution function $f(x, y)$ in $L^2(\mathbb{R}^2)$.*

The readers who have not learned functional analysis can skip this theorem, however, for readers' convenience, let us briefly explain what is claimed by Theorem 3.2.

$$L^2(\mathbb{R}^2) = \{\varphi(x, y) \mid \int_{\mathbb{R}^2} |\varphi(x, y)|^2 \, dx \, dy < \infty\} \tag{3.25}$$

is a linear space and for $\varphi \in L^2(\mathbb{R}^2)$ its norm is defined by

$$\|\varphi\|_{L^2(\mathbb{R}^2)} = \left(\int_{\mathbb{R}^2} |\varphi(x, y)|^2 \, dx \, dy \right)^{\frac{1}{2}}. \tag{3.26}$$

Theorem 3.2 claims that

$$\int_{\mathbb{R}^2} |f(x, y) - g_l(x, y)|^2 \, dx \, dy \tag{3.27}$$

is very small.

3.2 Three-Dimensional Shape Reconstruction of Asteroids

Today, asteroid explorations are being eagerly studied and performed. OSIRIS REx developed in the US and Hayabusa 2 developed in Japan are well known as asteroid probes. In asteroid exploration, if an asteroid is found, an attempt is made to reconstruct its 3D (three-dimensional) shape from the pictures taken by the asteroid probe. This problem is one of 3D shape reconstruction from 2D images or projections. In this section, we study a very simplified problem, where we assume that the asteroid does not revolve on its own axis and that the position of the asteroid probe is fixed. In the real case, the asteroid revolves on its own axis. It is also a very difficult problem to know the precise position of the asteroid probe, which would never be stationary in the real case.

For $\theta \in S^2, s \in \mathbb{R}$,

$$\begin{aligned} H(\theta, s) &:= \{\boldsymbol{x} \in \mathbb{R}^3\,;\ \boldsymbol{x} \cdot \theta = s\}, \\ \boldsymbol{x}_{(\theta,s)} &:= \boldsymbol{x} - (\boldsymbol{x} \cdot \theta - s)\theta, \end{aligned} \tag{3.28}$$

which represents the plane where the 2D picture is projected (see Fig. 3.2).

Let $\boldsymbol{x} = (x, y, z)$ and $\boldsymbol{x}_{(\boldsymbol{\theta}_i, s_i)} = (x_i, y_i, z_i)$ for $i = 1, \cdots, n$, where $\boldsymbol{\theta}_i = (\theta_{i1}, \theta_{i2}, \theta_{i3}) \in S^2$.

If we try to reconstruct the position $\boldsymbol{x}$ on the asteroid from $n > 3$ pictures then we have the following overdetermined system of $3n$ linear equations:

$$\begin{cases} x - (x\theta_{i1} + y\theta_{i2} + z\theta_{i3} - s_i)\theta_{i1} = x_i, \\ y - (x\theta_{i1} + y\theta_{i2} + z\theta_{i3} - s_i)\theta_{i2} = y_i, \\ z - (x\theta_{i1} + y\theta_{i2} + z\theta_{i3} - s_i)\theta_{i3} = z_i, \end{cases} \tag{3.29}$$

where $s_i = x_i\theta_{i1} + y_i\theta_{i2} + z_i\theta_{i3}$. This system (3.29) is equivalent of the following system:

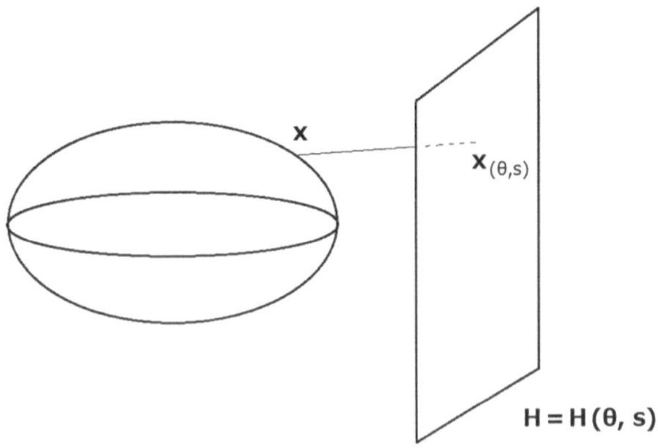

Fig. 3.2 3D reconstruction of an asteroid

$$\begin{cases} (1-\theta_{i1}^2)x - \theta_{i1}\theta_{i2}y - \theta_{i1}\theta_{i3}z = x_i - s_i\theta_{i1}, \\ -\theta_{i2}\theta_{i1}x + (1-\theta_{i2}^2)y - \theta_{i2}\theta_{i3}z = y_i - s_i\theta_{i2}, \\ -\theta_{i3}\theta_{i1}x - \theta_{i3}\theta_{i2}y + (1-\theta_{i3}^2)z = z_i - s_i\theta_{i3}, \end{cases} \tag{3.30}$$

which is a typical example of overdetermined system of linear equations.

The above systems (3.29) and (3.30) are obtained by simplification. The following problems are left for practical application.

Problem 3.6 In the real practice of 3D shape reconstruction of asteroids, we have the following problems:

- We have to take into account the revolution of the asteroid on its own axis.
- We can determine the direction the asteroid probe is facing in but it is almost impossible to know its precise position, by which we a priori know θ_i's but we do not know s_i's.
- We have to take into account that the asteroid probe is always moving by its own inertia force and by universal gravitation between the asteroid and the asteroid probe: determining its moving vector is a very difficult problem.

For more advanced research including some answers to Problem 3.6, cf. [10], where the analysis of a big system (2.12) with large n is necessary.

Remark 3.2 3D shape reconstruction from 2D images or projections is a very popular research topic in computer vision, whose objects are variety of things. 3D shape reconstruction of asteroids is one such example.

3.3 Precise Probing of Rebar Endpoints

In this section, we discuss how to detect the exact position of the rebar (reinforcing steel bar) endpoints in RC (reinforced concrete) structures by a non-destructive inspection. We note that although there are many studies on the detection of the rebar in RC structures non-destructively, for example, see [6, 24], there exists no non-destructive inspection technique to precisely probe the rebar.

There are advantages detecting the exact position of the rebar, among which, we introduce the most important two. The first one is that the knowledge of the exact position of the rebar enables us to use a non-destructive inspection technique for the concrete cover in RC structures by ultrasonic measurements, for which see [11, 20]. In the maintenance of RC structures, it is one of the most important tasks to keep the concrete cover in a sound state in order to prevent the rebar from getting corroded. The concrete cover is the layer of concrete between the surface of the structure and the first rebar. If we find cavities or nests inside the concrete cover, then we can shave the cover and refill the layer with new concrete or mortar. In this shaving, knowledge of the position of the rebar is of great help to avoid the damage to the rebar. If we know the precise position of the rebar we can shave the concrete cover

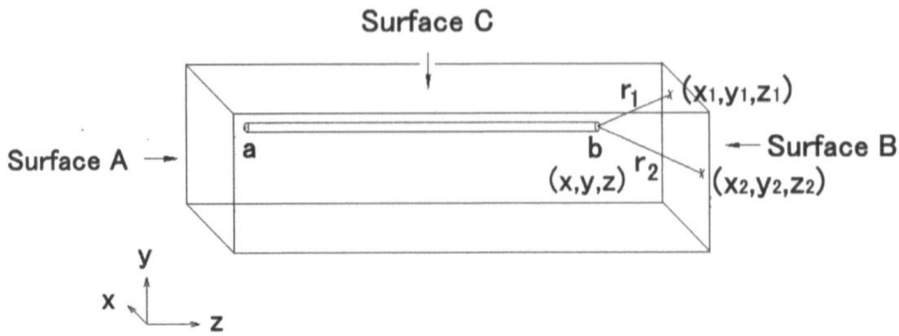

Fig. 3.3 A cuboid test piece

very closely to the rebar by machine. This is the second best advantage of detecting the position of the rebar. If we do not know the exact position, we have to shave the concrete cover by hand, which involves much labor cost and time.

In this section, as the first step, we develop a non-destructive inspection technique to exactly probe for the endpoints of the rebar, which is one of the most difficult problems in rebar detection.

Assume that a cuboid RC structure contains a rebar which is situated parallel or perpendicular to each edge surface. See Fig. 3.3.

In Fig. 3.3, we set the axes as shown. We reconstruct the endpoints of the rebar. For the reconstruction of the endpoint b in Fig. 3.3, we take suitable measurement points on the edge surface B and around it, and measure the thickness of concrete cover, equivalently the distance between the measurement points and the endpoint. We let the endpoint $b = (x, y, z)$ and take n measurement points $(x_1, y_1, z_1), \cdots, (x_n, y_n, z_n)$, whose distances from the endpoint b are measured as $r_1, \cdots, r_n$, respectively.

For the measurement, we apply the apparatus (cf. [16]) to measure by electromagnetic induction, the thickness of concrete cover for reinforcement. Its manual tells us to scan the surface with the apparatus parallel to the rebar for the measurement of the thickness of the concrete cover.

It is known that the measurements by this apparatus are not accurate and it is not suitable for detecting the rebar endpoints. In fact, for the time being, there exists no suitable apparatus for our purpose to exactly probe for the rebar endpoints. Therefore, it is necessary to develop a new technology in order to obtain more accurate measurements of the rebar endpoints using an apparatus including some large errors. For the development of our new technology, we used this apparatus in another way than the manual tells us. We mainly scanned the surface A or B (in Fig. 3.3) using this apparatus to measure the distance between the endpoint of the rebar (the point b in Fig. 3.3) and the measurement points where we put the probe (on the surface B in Fig. 3.3). In these measurements, we did not scan the surface parallel to the rebar. We modified the inaccuracy of the measurements by application of Theorem 2.1.

For the reconstruction of the endpoint b in Fig. 3.3, we took measurement points on the edge surface B and around it (for example, some points on the surface C which were near to the surface B) and measured the distances between the endpoint (the point b in Fig. 3.3) and the measurement points. We let the endpoint $b = (x, y, z)$ and took $n > 3$ measurement points $(x_1, y_1, z_1), \cdots, (x_n, y_n, z_n)$, whose distances from the endpoint b were measured as $r_1, \cdots, r_n$, respectively.

By these measurements, we obtained the overdetermined system (1.2) of quadratic equations; however, it is well known that it is impossible to measure the thickness of concrete cover accurately using the existing devices. Henceforth all measurements necessarily contained a few large errors and some noise. Therefore, we had to construct an approximate solution of the system (1.3) containing such inaccuracies. We note that Assumption 2.1 is satisfied since if all measurements are precisely accurate then there exists a solution (the endpoint of the rebar) to (1.2). We also note that Assumption 2.3 is also satisfied by taking suitable and suitably many measurement points. By subtracting the j-th equation from the i-th one of the system (1.3) for $i < j$ (the semi-equivalent deformation (iii)) and dividing both sides of the differences by 2 (the semi-equivalent deformation (ii)), we obtained the overdetermined system of $\frac{n(n-1)}{2}$ linear equations,:

$$(x_i - x_j)x + (y_i - y_j)y + (z_i - z_j)z = \frac{1}{2}(x_j^2 + y_j^2 + z_j^2 - x_i^2 - y_i^2 - z_i^2 + r_i^2 - r_j^2), \tag{3.31}$$

where we have also applied the semi-equivalent deformation (i). Note that the system (3.31) is an overdetermined system of linear equations.

Therefore, we could apply our main theorem, Theorem 2.1, to construct an approximate solution.

The author's group verified by experiments using real RC structures that the reconstruction technique established in the previous chapter works very well for the reconstruction of the x- and the y-coordinates of the rebar endpoint with $6 \leq n \leq 10$, but the z-coordinate could not be precisely reconstructed however large number n of measurements. A solution to precisely reconstruct the z-coordinate is given in [19].

We note that it is sufficient to precisely reconstruct the x- and the y-coordinates of the rebar endpoint to modify the measurements of the thickness of the concrete cover, since the comparison of the x- and the y-coordinates of the rebar endpoint and the real measured value of the thickness of the concrete cover near the rebar endpoint would give how to modify the measured of the thickness of the concrete cover.

The author's group is developing new non-destructive inspection devices for RC structures applying this idea to accurately probe of the rebar endpoints, a part of which is introduced in [19].

3.4 GNSS Positioning Technique

GNSS stands for global navigation satellite system. Developments of this system are also known as QZSS (Japan), GPS (the US), GLONASS (Russia) and Galileo (EU). GNSS is an essential infrastructure that supports various fields, including surveying, car navigation, precision farming, search and rescue. It consists of navigation satellites orbiting around the earth that constantly broadcast navigation signals. A GNSS user on the ground employs a receiver to analyze the signals to determine her or his position (see Fig. 3.4).

For general theory of GNSS positioning technique, we refer the readers to [26], for example.

The mathematical model for GNSS positioning technique is similar to the one for the exact probing rebar endpoints in the previous section. Signals from satellites carry time stamps which represent signal transmitting times generated with the precise atomic clocks onboard the satellites. A receiver analyzes the difference between transmitting times and receiving times that gives the distance between the satellite and the receiver. In general, the locations of navigation satellites are known whose information is also sent in the signals from the satellites. The clocks of receivers have some offsets and they are unknown. So, in the simplest case, there are four unknown parameters (three parameters for a receiver's position, one parameter for a clock offset), and at least four satellites are required to determine a receiver's location. A receiver usually takes signals from 6 to 10 satellites at a time, which means 6 to 10 geometrical constraints are available to determine the receiver's location.

If we can determine the time offset, then we have an overdetermined system (1.2) of quadratic equations with errors and noise, the same one as for the exact probing rebar endpoints. It is interesting that the number of measurements (6 to 10) are also almost the same as for the rebar endpoints. The measurements in GNSS are much

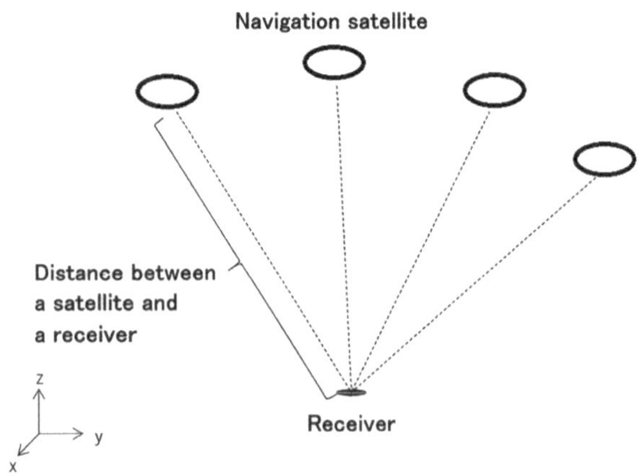

Fig. 3.4 Concept of GNSS positioning

more accurate than in rebar endpoint detection; however, they must contain some errors because:

- The velocity of the microwave signal changes in accordance with the condition of the air (pressure, humidity, and so on) and ionosphere.
- The inner atomic clock of the satellite gains or loses a slight amount of time.

These two causes are the main ones for the measurements in GNSS to be inaccurate. For the time being, the more accurate GNSS devices receive more information about the air condition and so on, which makes the devices bigger and more expensive. The author claims that, instead of receiving much information, the application of the idea of the least-square solution to the semi-equivalent system (3.31) is of great help. It makes the device much cheaper and smaller, and the positioning is sufficiently accurate. In this new positioning technique, we have to carefully study the characteristics of the accuracy of the positioning of each coordinate. We know by experience that the positioning of the z-coordinate is not more accurate than that of x- and y-coordinates. Therefore, for GNSS positioning, some special method for the positioning of the z-axis may be required. The new idea for GNSS positioning technique, including the determination of the time offset and precise positioning of the z-axis, is under investigation.

Remark 3.3 In the conventional GNSS positioning, the following system

$$\begin{cases} r_1 = \sqrt{(x - x_1)^2 + (y - y_1)^2 + (z - z_1)^2} + \delta_1 \\ \qquad \cdots \\ r_m = \sqrt{(x - x_m)^2 + (y - y_m)^2 + (z - z_m)^2} + \delta_m \end{cases} \tag{3.32}$$

equivalent to the system (1.3), where δ_i are errors, is often applied in the analysis of the receiver's position (cf. [9]). In the analysis of (3.32), they apply asymptotic expansion of each equation in the system (3.32), which must contain the additional errors in the truncation of the expansion. The author claims that the analysis of the system (1.3), equivalently its semi-equivalent system (3.31), is the easier, the more precise and the less costly, by virtue of the discussion in the previous chapter. Therefore, we have posed a new technique of GNSS positioning technique by application of semi-equivalent systems.

3.5 A Remark on This Chapter

In the course of elementary linear algebra, it is a must to learn how to solve an underdetermined linear system, which is very important to learn the structure of affine spaces. However, it is rarely taught how to analyze an overdetermined system of linear equations with no solution in the course of elementary linear algebra, which is often required in practical applications, as we have seen in Sects. 3.1–3.4. The author's group [14] claims that it is very important and interesting to teach how to analyze an overdetermined system of linear equations with no solution in the course of elementary linear algebra. In the paper [14], the authors discuss the importance of teaching how to analyze an overdetermined system of linear equations with no solution in the course of elementary linear algebra from two viewpoints: one is the teaching of elementary mathematics not limited to linear algebra and the other is practical applications which can raise the students' motivation to study mathematics. In the same paper, the authors also propose a variety of teaching materials of the overdetermined systems of linear equations with no solution in accordance with the degree of advancement of the classes.

Exercises

Exercise 3.1

(a) Prove Proposition 3.1.
(b) Prove Theorem 3.1.
(c) Solve (3.11) for $u(t)$.

Exercise 3.2 Derive (3.14) from (3.13).

Exercise 3.3 (Uniqueness of the Radon transform)

(a) Prove that $Rf \equiv 0$ implies that $f \equiv 0$ for $f \in L^1(\mathbb{R}^2)$.
(b) Show that the answer to Problem 3.3 is affirmative, that is, $Rf_1 \equiv Rf_2$ implies that $f_1 = f_2$ for $f_1, f_2 \in L^1 \cap C(\mathbb{R}^2)$.
(c) Show that $Rf \equiv 0$ implies that $f \equiv 0$ for $f \in L^1(\mathbb{R}^n)$ for $\mathbb{N} \ni n \geq 3$.
(d) Show that the answer to Problem 3.3 is affirmative, that is, $Rf_1 \equiv Rf_2$ implies that $f_1 = f_2$ for $f_1, f_2 \in L^1 \cap C(\mathbb{R}^3)$, for $\mathbb{N} \ni n \geq 3$.

In order to solve Problem 3.3, knowledge of the elements of Lebesgue integral [2, 25] and the Fourier analysis [18] is required. The readers who have learned neither Lebesgue integrals nor Fourier analysis may skip this problem.

Exercise 3.4

(a) Show that Problem 3.1 is reduced to Problem 3.5 by G. N. Hounsfield's idea.
(b) Show Theorem 3.2.

Exercise 3.5 Consider the following system:

$$\begin{cases} (x-1)^2 + (y-1)^2 = \rho^2, \\ (x+1)^2 + (y-1)^2 = \rho^2, \\ (x+1)^2 + (y+1)^2 = \rho^2, \\ (x-1)^2 + (y+1)^2 = \rho^2, \end{cases} \tag{3.33}$$

for any $\rho > 0$.

(a) Find the value $\rho = \rho_0$ for which the solution of the system (3.33) is unique and give the unique solution.

(b) Prove that the system (3.33) for $\rho \neq \rho_0$ would allow no solution.

(c) Derive a semi-equivalent system of linear equations to the system (3.33) for any $\rho > 0$.

(d) Give the unique least-square solution to the system (3.33) for any $\rho > 0$ by application of the semi-equivalent system of linear equations obtained in (c).

(e) Prove that the unique least-square solution to the system (3.33) for any $\rho > 0$ coincides with the unique solution for $\rho = \rho_0$.

Chapter 4
Structure of the Accuracy and the Inaccuracy of the Least-Square Solutions

Abstract In this chapter, we study the structure of the accuracy and the inaccuracy of the least-square solutions. We also study how the inaccuracy of the least-square solution happens and how to treat it.

In this chapter, we study the structure of the accuracy and the inaccuracy of the least-square solutions, for the case of small (about 6–10) measurement data in the system (1.3) where we approximately reconstruct the object point (x, y, z) from the measurements $\rho_1, \cdots, \rho_m$ of the distances between the object (x, y, z) and the measurement points (x_j, y_j, z_j), $j = 1, 2, \cdots, m, 6 \leq m \leq 10$.

4.1 General Settings

In the examples introduced in Sects. 3.3 and 3.4, the reconstructions of the x- and y-coordinates by application of the least-square solution to the semi-equivalent overdetermined system of linear equations are very accurate using from 6 to 10 measurement data. On the other hand, the reconstruction of the z-coordinate would not be accurate, however many measurement data are given. In this section, we study why accuracy for the x- and y-coordinates and inaccuracy for the z-coordinate happens.

Consider the circle $x^2 + y^2 = 2$, for a very typical example. If we try to reconstruct its center as the least-square solution from the four points $(\pm 1, 0)$, $(0, \pm 1)$ with the distance $\rho > 0$ between the origin and these four points, then, for any value $\rho > 0$, we obtain the exact point, the origin, as the least-square solution. That is, if we try to give the least-square solution to the system (2.7) for any $\rho > 0$, then it is given as $x = y = 0$. Therefore, if we take observation points from all directions (in the angle 2π) around the object (in this example, the object is $(x, y) = (0, 0)$) then the object is accurately reconstructed by application of the least-square solution with not so many observation data. See Fig. 4.1, even if the all measurements are larger or smaller than the real values.

T. Takiguchi, *The Least Square Solution to an Overdetermined System*, SpringerBriefs in Mathematical Methods, https://doi.org/10.1007/978-981-92-0064-1_4

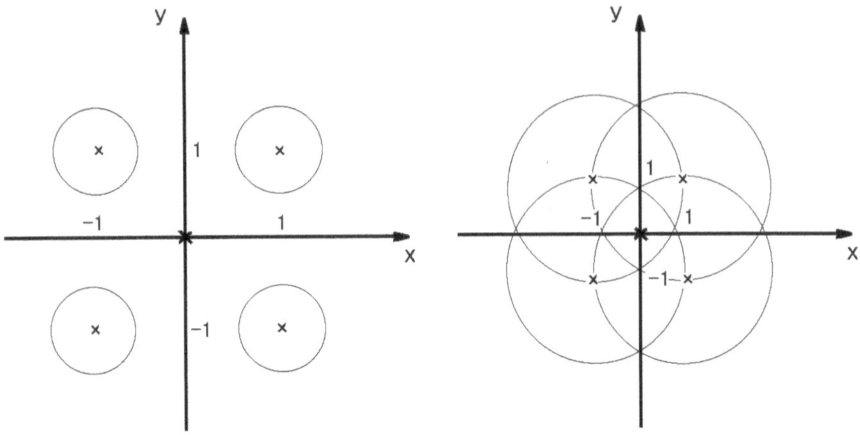

Fig. 4.1 Reconstruction of the origin

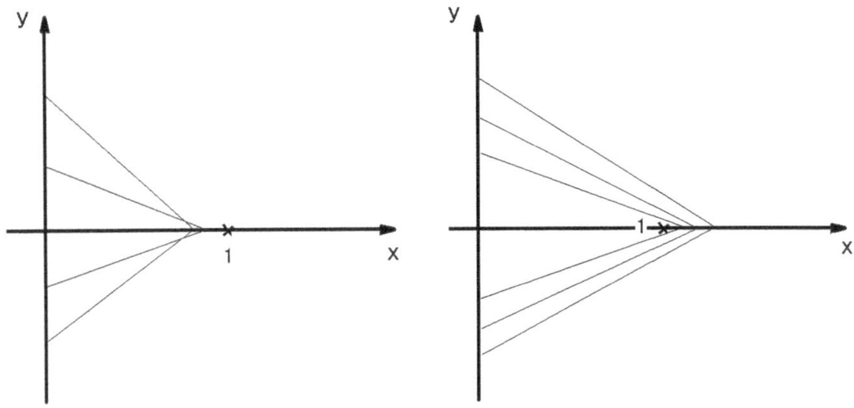

Fig. 4.2 Reconstruction of the point (1, 0)

On the other hand, consider the case where we try to reconstruct the point $C = (1, 0)$ in the plane from the points $D_k = (0, \frac{k}{10})$, $k = -10, -9, \cdots, -1, 0, 1, \cdots, 9, 10$, where the distance between C and D_k is given by $d(C, D_k) = \sqrt{100 + \frac{k^2}{100}}$. If we take all the distances between C and D_k smaller than the real one, $d(C, D_k) = \sqrt{100 + \frac{k^2}{100}}$ then the least-square solution gives the coordinates of the point C as $(x, 0)$, where $x < 1$ (Fig. 4.2, left). If we take all the distances between C and D_k larger than the real one, $d(C, D_k) = \sqrt{100 + \frac{k^2}{100}}$, then the least-square solution gives the coordinates of the point C as $(x, 0)$, where $x > 1$ (Fig. 4.2, right). Therefore, if we cannot take the observation points on only one side from the object (in the angle $< \pi$) in the plane, then the reconstruction of the object cannot be accurate, however many observation data we take.

Let us study the structure of the errors in the examples introduced in Sects. 3.3 and 3.4.

In Sect. 3.3, we reconstructed the rebar endpoints by application of the apparatus by electromagnetic induction to measure the thickness of concrete cover for reinforcement. As was mentioned in Sect. 3.3, the measurements by this apparatus are known to be inaccurate. By the experiments, the author's group [19] proved that the errors in the measurements by this apparatus depend on the structure of the concrete; if the concrete is very solid, where limestone is included as gravel and more cement is included, then all the measurements of the distances by this apparatus are longer than the real ones, if the concrete is not solid, then all the measurements of the distances by this apparatus are shorter than the real ones. The author's group [19] also verified that the errors in measurements are proportional to the lengths of the real distances to be measured.

We have a similar problem in GNSS positioning introduced in Sect. 3.4. In GNSS positioning, a satellite projects a microwave signal with the information of the location of the satellite and the time stamp when the microwave was projected. Therefore the receiver can determine the distance between the satellite and the receiver obtained by

$$vt, \tag{4.1}$$

where v is the velocity of the microwave and t is the time the microwave takes from the satellite to the receiver. However, v in (4.1) changes a little in accordance with the weather in the area where the satellites and the receiver are, which is the main cause of the errors in the measurements in t and the distances in GNSS positioning. When such error in t happens, we have either of the following two cases: In one case, all the measured travel times t are longer than expected. In the other case, all the measured travel times t are shorter than expected where the air is very dry. If t becomes longer, then all measurements of the distances are longer than the real ones where the air is humid. If t becomes shorter, then all measurements of the distances are shorter than the real ones where the air is dry. There are other causes of change in the velocity of the microwave, such as the condition of the ionosphere and/or the troposphere. Here we assume that the humidity is the main reason for a change in the velocity of the microwave.

In view of the above discussion, we can conclude that the following property holds.

Property 4.1 In the examples introduced in Sects. 3.3 and 3.4, either of the following two cases happens: In one case, all measurements of the distances are longer than the real ones. In the other case, all measurements of the distances are shorter than the real ones. The errors in the measurements are almost proportional to the lengths of the real distances to be measured.

By Property 4.1, we can say that the errors in the measurements are stable in the sense that they are proportional to the real distances to be measured. We note that other very small errors in several causes are also included in the errors in the measurements.

As we have seen in the examples in Figs. 4.1 and 4.2, Property 4.1 is very unfavorable for the determination of the z-coordinate of the endpoints and it cannot be improved by increasing the number of measurements. On the other hand, the stability in errors of the measurement is very favorable for the reconstruction of the x- and y-coordinates. For their reconstruction, very accurate approximations are obtained from only six to eight observation data. In this chapter, we shall study this accuracy and inaccuracy. In the rest of this chapter, unless otherwise mentioned, we assume Property 4.1.

4.2 The Stable Problems

In this section, we study the structure of the accuracy in the reconstruction of the x- and y-coordinates in the examples introduced in Sects. 3.3 and 3.4.

Definition 4.1 Let us consider the following system:

$$\begin{cases} (x - x_1)^2 + (y - y_1)^2 + (z - z_1)^2 = (1 + \varepsilon)^2 r_1^2 + \varepsilon_1, \\ \qquad \cdots \\ (x - x_m)^2 + (y - y_m)^2 + (z - z_m)^2 = (1 + \varepsilon)^2 r_m^2 + \varepsilon_m, \end{cases} \tag{4.2}$$

where r_j's are real distances, $|\varepsilon_j|$'s are very small and $|\varepsilon| < \frac{1}{10}$. If we can take measurement points (x_j, y_j, z_j)'s from all the eight octants in the orthogonal coordinate system in $\mathbb{R}^3$ with the object point (x, y, z) as its origin, then we call the problem to reconstruct the object (x, y, z) in the system (4.2) *a three-dimensional stable problem.*

From the viewpoint of Property 4.1, the right-hand sides of the system (4.2) are given, where ε_j's are small errors. The condition $|\varepsilon| < \frac{1}{10}$ is given by the experience of the measurements in the problems introduced in Sects. 3.3 and 3.4. In the problem of GNSS, $|\varepsilon|$ can be taken much smaller.

Definition 4.2 Let us consider the following system:

$$\begin{cases} (x - x_1)^2 + (y - y_1)^2 = (1 + \varepsilon)^2 r_1^2 + \varepsilon_1, \\ \qquad \cdots \\ (x - x_m)^2 + (y - y_m)^2 = (1 + \varepsilon)^2 r_m^2 + \varepsilon_m, \end{cases} \tag{4.3}$$

where r_j's are real distances, $|\varepsilon_j|$'s are very small and $|\varepsilon| < \frac{1}{10}$. If we can take six measurement points (x_j, y_j)'s $(j = 1, 2, 3, 4, 5, 6)$ whose arguments θ_j's satisfy $\frac{(j-i)\pi}{3} < \theta_j < \frac{j\pi}{3}$ in the orthogonal coordinate system with the object point (x, y)

as its center, then we call the problem to reconstruct the object (x, y) in the system (4.3) *a two-dimensional stable problem.*

We say that the problems to reconstruct the x- and y-coordinates in the examples introduced in Sects. 3.3 and 3.4 are stable in the sense of Definition 4.2.

4.3 The Unstable Problems

In this section, we study the structure of the inaccuracy in the reconstruction of the z-coordinate in the examples introduced in Sects. 3.3 and 3.4.

Definition 4.3 In the system (1.3), if we can only take all measurement points (x_j, y_j, z_j)'s in the half space $\{z > 0\}$ or $\{z < 0\}$, then we call the problem to reconstruct the z-coordinate of the object (x, y, z) in the system (1.3) *an unstable problem to reconstruct the z-coordinate.*

Definition 4.4 In the system (1.3), if we can only take all measurement points (x_j, y_j, z_j)'s in the half space $\{z > 0\}$ or $\{z < 0\}$, and if we can take six measurement points $(r_j \cos\theta_j, r_j \sin\theta_j, z_j)$'s $(j = 1, 2, 3, 4, 5, 6)$, where $r_j > 0$ and $\frac{(j-i)\pi}{3} < \theta_j < \frac{j\pi}{3}$, then we call the problem to reconstruct the object (x, y, z) in the system (1.3) *stable in the x- and y- coordinates and unstable in the z-coordinate.*

Precisely speaking, the following proposition holds.

Proposition 4.1 *The problems introduced in Sects. 3.3 and 3.4 are stable in the x- and y- coordinates and unstable in the z-coordinate.*

The instability in the reconstruction of the z-coordinate is almost trivial by the example shown in Fig. 4.2. We know by experience that the problems introduced in Sects. 3.3 and 3.4 are stable in x- and y- coordinates; however, its theoretical proof is left open.

In the rest of this section, let us study how to show the first part of Proposition 4.1, which shall also play an important role to correct the instability in the reconstruction of the z-coordinate in the next section. The answer to the following problem shows the first part of Proposition 4.1.

Problem 4.1 Prove that the x- and y-coordinates of the least-square solution to the system (4.2) are very close to the counterparts of the real solution (x, y, z).

The following problem is also interesting in the study of Proposition 4.1.

Problem 4.2 Study the connection of the problem to reconstruct the x- and y-coordinates of the solution to the system (4.2) and the solution to the system (4.3).

4.4 Correction of the Instability in the Reconstruction of the z-coordinate

As we have already seen, reconstruction of the z-coordinate is unstable in the problems introduced in Sects. 3.3 and 3.4. In this section, we study a method to correct this instability by application of the stability is the reconstruction of the x- and y-coordinates. Since this is a little difficult and too specialized, the readers who are not interested can skip this section.

In this section, for simplicity, we assume the following assumption.

Assumption 4.1 In the system (4.2), we assume the following in this section:

(i) The object to be reconstructed is $(x, y, z) = (0, 0, 0)$.
(ii) $z_j > 0$ for any $1 \leq j \leq m$.

By Assumption 4.1 (i), the problem to reconstruct the object (x, y, z) in the system (4.2) is stable in x- and y- coordinates and unstable in the z-coordinate.

By the discussions in Sects. 4.2 and 4.3, we obtain a good approximation $(x, y) \approx (0, 0)$ as the least-square solution to the system (4.2). Note that the right-hand side of each Eq. (4.2) is obtained by measurement. We substitute this approximate solution $(x, y) \approx (0, 0)$ in (4.2) and obtain

$$\begin{cases} \delta_1^2 + d_1^2 = (1+\varepsilon)^2 r_1^2 + \varepsilon_1 + \epsilon_1, \\ \quad \cdots \\ \delta_m^2 + d_m^2 = (1+\varepsilon)^2 r_m^2 + \varepsilon_m + \epsilon_m, \end{cases} \tag{4.4}$$

where $\delta_j^2 = x_j^2 + y_j^2$ are known, ϵ_j are small errors. By (4.4), we obtain

$$\begin{cases} (1+\varepsilon)^2 r_1^2 = z_1^2 - \epsilon_1 - \varepsilon_1 - \delta_1^2 \\ \quad \cdots \\ (1+\varepsilon)^2 r_m^2 = z_m^2 - \epsilon_m - \varepsilon_m - \delta_m^2 \end{cases} \tag{4.5}$$

where r_j's and ε are unknowns. If we know ε then we can correct the system (4.2). If we divide all the right hand sides by $(1+\varepsilon)^2$, then we obtain the closer value to the square r_j^2 of the real length r_j. Giving the least-square solution of the system (4.2), all right-hand sides of whose equations are modified as above, we obtain a good approximation of the z-coordinate of the object. Therefore, the following problem is posed.

Problem 4.3 Give a least-square solution of ε in the system (4.5).

It is sufficient to give approximate values of r_j and ε, since we can approximate $-\epsilon_j - \varepsilon_j - \delta_j^2$ in the j-th equation in (4.5) as 0. Therefore, Problem 4.3 is approximately solved as the minimizer of

$$\varphi(\varepsilon) = \sum_{j=1}^{m} \left((1+\varepsilon)^2 r_j^2 - z_j^2 - \delta_j^2 \right), \tag{4.6}$$

which is given by the value $\varepsilon_0 = \varepsilon_0(r_1, \cdots, r_m)$ satisfying $\varphi'(\varepsilon_0) = 0$. We have to solve the system

$$(1+\varepsilon)^2 r_j^2 = \rho_j^2 \quad (j = 1, 2, \cdots, m), \tag{4.7}$$

where ρ_j are the measured distances, in order to obtain the concrete approximate values of the right-hand sides in the system (4.2).

Although the above method gives a good approximation of ε which yields a good approximation of the z-coordinates, in practical applications, there are other practical solutions of the instability in the z-coordinate.

In the detection of the endpoints in RC structures introduced in Sect. 3.3, we can take observation points in the manner that the all the z-coordinates z_j in the observation points (x_j, y_j, z_j) are the same. Therefore, in the system (4.5), $z_1 = \cdots = z_m = d$ holds. Hence it is sufficient to give the minimizer of

$$\varphi(\varepsilon) = \sum_{j=1}^{m} \left((1+\varepsilon)^2 r_j^2 - d^2 - \delta_j^2 \right), \tag{4.8}$$

In GNSS positioning introduced in Sect. 3.4, we can obtain the heights z_j of the satellites and $|\varepsilon|$ can be taken very small. Therefore, we can assume:

Assumption 4.2 In GNSS positioning, we can assume the followings.

(i) The heights z_j of the satellites are known.
(ii) In the system (4.6), ε^2 can be regarded as 0.

In this case, it is sufficient to give the minimizer of

$$\varphi(\varepsilon) = \sum_{j=1}^{m} \left((1+2\varepsilon) r_j^2 - z_j^2 - \delta_j^2 \right), \tag{4.9}$$

where z_j are known.

Remark 4.1 If we try to correct the instability in the reconstruction of the z-coordinate, then mathematical problems becomes little difficult; however, in many practical applications, there are other engineering methods to approximately obtain r_j and/or ε than mathematical ones. For example, cf. [19] for rebar endpoint detection.

Exercises

The exercises in this chapter are difficult and include open problems. If you are interested in the research of the problems in least-square solutions, then please try them. Otherwise, please skip the problems other than Exercise 4.2.

The solutions to the problems in this chapter are omitted.

Exercise 4.1 Answer the following open problems:

(a) Solve Problem 4.1.
(b) Solve Problem 4.2.

Exercise 4.2 Solve Problem 4.3.

Exercise 4.3 Answer the following open problems:

(a) Give the minimizer of (4.8).
(b) Give the minimizer of (4.9).

Chapter 5
A Method to Construct an Approximate Solution to an Overdetermined System

Abstract In this chapter, we study how to construct an approximate solution to an overdetermined system. We study two cases; one is how to construct an approximate solution to a large or huge overdetermined system, the other is the case where iteration to construct an approximate solution to an overdetermined system is necessary.

In this chapter, we study how to construct an approximate solution to a large or huge overdetermined system of linear equation (2.12). We also treat the case where iteration to construct an approximate solution to an overdetermined system is necessary.

5.1 The Main Problem in This Chapter

Our original problem is to give an approximate solution to the overdetermined system (1.4), containing errors and noise, obtained by measurements. By Assumption 2.2, the system (2.12) is derived as a semi-equivalent system of the system (1.4) and by Assumption 2.3, $\mathrm{rank}(A) = n$ holds for the coefficient matrix A in (2.12). Therefore, by virtue of Theorem 2.1, our original problem is reduced to giving an approximate solution to the system (2.12).

Problem 5.1 Give an approximate solution to the system (2.12).

Theoretically, this problem has been solved by Theorem 2.1, by giving the unique solution to the system (2.28). By (2.28), the least square solution to (2.12) is given by (2.29), where we note that ${}^tAA$ is regular by Assumption 2.3 and Proposition 2.5.

The numerical implement of (2.29) for practical application is not difficult if n is not so large, for example $n \leq 10$ in the examples introduced in Sects. 3.3 and 3.4. It is, however, not easy if n is large, for example, in the problem of CT introduced in Sect. 3.1 and a more generalized 3D reconstruction of an asteroid by pictures taken by an asteroid probe, mentioned at the end of Sect. 3.2 (see also [10]).

T. Takiguchi, *The Least Square Solution to an Overdetermined System*, SpringerBriefs in Mathematical Methods, https://doi.org/10.1007/978-981-92-0064-1_5

Though computation of ${}^tA$ is not difficult however large the number n is, it is costly and difficult to compute $({}^tAA)^{-1}$ for large or huge n.

Therefore, the main problem in this chapter is to study how to give an approximate solution to the system (2.12) for large or huge n in an easily and cheaply computable way by not so high-spec computers.

Problem 5.2 Give an approximate solution to the system (2.12) for large or huge n, in an easily and cheaply numerically computable way, without application of Theorem 2.1.

This problem shall be studied in the next section. We have another problem. In some problems, we have to repeatedly apply Theorem 2.1. In each step to calculate $({}^tAA)^{-1}$, numerical errors accumulate even if the size of $({}^tAA)$ is not so large. Therefore, for the case where the iteration of the calculus to give a least square solution to an overdetermined system, it is not suitable to apply Theorem 2.1 to calculate the least square solution. In Sect. 5.3, we shall discuss this problem.

5.2 Treatment of the Large or Huge System

In this section, we give an answer to Problem 5.2.

Problem 5.2 was studied by G. N. Hounsfield for the practicalization of CT (cf. [22] in Chapter 2). His idea was what is now called the ART (algebraic reconstruction technique) or Kaczmarz method. He hit upon this idea all by himself without knowing the ART or Kaczmarz method (for these methods, see [13]).

We first note that the i-th equation in (2.12),

$$a_{i1}x_1 + a_{i2}x_2 + \cdots + a_{in}x_n = \widetilde{s_i}, \tag{5.1}$$

represents the hyperplane $H_i \subset \mathbb{R}^n$ whose normal vector is $\boldsymbol{a}_i := (a_{i1}, a_{i2}, \cdots, a_{in})$. Take any $\boldsymbol{x}_0 \in \mathbb{R}^n$ and define the sequence $\{\boldsymbol{x}_k\}$ by

$$\boldsymbol{x}_M := P\boldsymbol{x}_{M-1} \equiv P_k P_{k-1} \cdots P_1 \boldsymbol{x}_{M-1}, \; M = 1, 2, \cdots \tag{5.2}$$

where

$$P_i \boldsymbol{x} := \boldsymbol{x} - \frac{\boldsymbol{a}_i \cdot \boldsymbol{x} - s_i}{\|\boldsymbol{a}_i\|} \frac{\boldsymbol{a}_i}{\|\boldsymbol{a}_i\|} \tag{5.3}$$

is the orthogonal projection of $\boldsymbol{x}$ onto the hyperplane H_i. If the system (2.12) has a unique solution, then it is easily proved that the limit

$$\boldsymbol{x}_\infty = \lim_{M \to \infty} \boldsymbol{x}_M \tag{5.4}$$

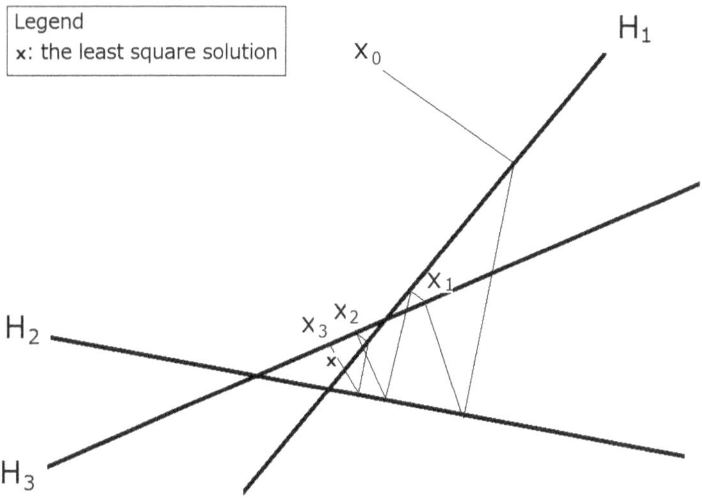

Fig. 5.1 Iterated approximation and the least square solution

gives the unique solution to (2.12), which is called the ART or Kaczmarz method. Of course, the system has no solution in practice, because of errors and noise, all of which are small. Therefore, it can be said that the system (2.12) is *close to having a unique solution*. Therefore, (5.4) would give an approximate solution to (2.12), which works very well in practice. It is known that this iteration method almost converges in a few steps, $k = 3$ or 4 in (5.4), independently of the size of the system (cf. Fig. 5.1). We also note that if the errors in each equation in the system (2.12) are very small, then an approximate solution to (2.12) constructed by this iteration method and the least square solution are very close, both of which approximate what is required, see Fig. 5.1.

Remark 5.1 As we have mentioned above, $\boldsymbol{x}_\infty$ defined in (5.4) is an approximation of the least square solution to (2.12). However, in general, the average

$$\frac{1}{k}\sum_{j=1}^{k}\boldsymbol{x}^j, \tag{5.5}$$

where $\boldsymbol{x}^j = P_j P_{j-1}\cdots P_1\boldsymbol{x}_\infty$ for $j = 1, 2, \cdots, k$, gives a more approximate solution to the least square solution to (2.12). In the real numerical implementation of CT, $\boldsymbol{x}_4$ ($M = 4$) is almost the limit, that is, $\boldsymbol{x}_4 \approx \boldsymbol{x}_\infty$. In this case the average $\frac{1}{k}\sum_{j=1}^{k}\boldsymbol{x}_4^j$ with $\boldsymbol{x}_4^j = P_j P_{j-1}\cdots P_1\boldsymbol{x}_4$ for $j = 1, 2, \cdots, k$, is generally a better approximation than $\boldsymbol{x}_4$. See Exercises 5.2 and 5.3 below.

There are studies of such overdetermined systems of linear equations, cf. [3, 23] for some of them.

5.3 The Case Where the Iteration is Necessary

As mentioned in Sect. 5.1, in some problems, we have to repeatedly apply the solution to the overdetermined system of linear equations. As an example of such a problem, we introduce a problem to reconstruct the three-dimensional animation of an asteroid from the pictures taken by an asteroid probe [10].

If we apply Theorem 2.1 to such a problem, in each step to calculate $({}^tAA)^{-1}$, numerical errors accumulate even if the size of $({}^tAA)$ is not so large. Therefore, for the case where the iteration of the calculus gives the least square solution to an overdetermined system, it is not suitable to apply Theorem 2.1 to calculate the least square solution.

We apply a method similar to that of Sect. 5.2 in order to reduce the accumulation of the numerical errors. We first divide both sides of the j-th equation in (2.12) by

$$\|\boldsymbol{a}_j\| = \sqrt{a_{j1}^2 + a_{j2}^2 + \cdots + a_{jn}^2} \tag{5.6}$$

and obtain

$$\begin{cases} \alpha_{11}x_1 + \alpha_{12}x_2 + \cdots + \alpha_{1n}x_n = S_1, \\ \alpha_{21}x_1 + \alpha_{22}x_2 + \cdots + \alpha_{2n}x_n = S_2, \\ \quad \cdots\cdots\cdots \\ \alpha_{k1}x_1 + \alpha_{k2}x_2 + \cdots + \alpha_{kn}x_n = S_k, \end{cases} \tag{5.7}$$

where $S_j = \frac{s_j + \varepsilon_j}{\|\boldsymbol{a}_j\|}$.

Take any $\boldsymbol{x}_0 \in \mathbb{R}^n$ and define the sequence $\{\boldsymbol{x}_M\}$ by

$$\boldsymbol{x}_M := P\boldsymbol{x}_{M-1} \equiv P_k P_{k-1} \cdots P_1 \boldsymbol{x}_{M-1}, \ M = 1, 2, \cdots, \tag{5.8}$$

where

$$P_i \boldsymbol{x} := \boldsymbol{x} - (\boldsymbol{\alpha}_i \cdot \boldsymbol{x} - S_i)\boldsymbol{\alpha}_i \tag{5.9}$$

is the orthogonal projection of $\boldsymbol{x}$ onto the hyperplane H_i defined by the j-th equation in the system (5.7). We have modified the form of the orthogonal projections (5.3) to (5.9) in order to reduce the accumulation of numerical errors by division in the iteration. Usually, $M = 3$ or 4 gives a good approximation of the least square solution $\boldsymbol{x}_\infty = \lim_{M\to\infty} \boldsymbol{x}_M$ to the system (2.12); however, we apply

$$\widetilde{\boldsymbol{x}_l} = \frac{\boldsymbol{x}_l^1 + \boldsymbol{x}_l^2 + \cdots + \boldsymbol{x}_l^k}{k}, \tag{5.10}$$

for some $l \geq 3$ as a better approximation of the least square solution to the system (2.12) than $\boldsymbol{x}_l$, for whose reason see Exercise 5.2 (d) below. In the definition (5.10) of $\widetilde{\boldsymbol{x}_l}$, we note that

$$\boldsymbol{x}^j := P_j P_{j-1} \cdots P_2 P_1 \boldsymbol{x}. \tag{5.11}$$

In general the following proposition holds.

Proposition 5.1 *In any octant in $\mathbb{R}^n$, there exists a normal vector*

$$\boldsymbol{\alpha}_j = (\alpha_{j1}, \alpha_{j2}, \cdots, \alpha_{jk}) \tag{5.12}$$

in the j-th equation in the system (5.9). Let $\boldsymbol{x}_\infty = \lim_{M\to\infty} \boldsymbol{x}_M$ and

$$\widetilde{\boldsymbol{x}_\infty} = \frac{\boldsymbol{x}_\infty^1 + \boldsymbol{x}_\infty^2 + \cdots + \boldsymbol{x}_\infty^k}{k}. \tag{5.13}$$

See (5.11) for the definitions of $\boldsymbol{x}_\infty^j$'s. In this case, $\widetilde{\boldsymbol{x}_\infty}$ in (5.13) is a better approximation of the least square solution to the system (2.12) than $\boldsymbol{x}_\infty$.

The assumption of Proposition 5.1 is satisfied in practical CT, where we project very many X-rays at the human body. The proof of this proposition is almost trivial by Fig. 5.1.

Exercises

Exercise 5.1 Prove that (5.3) is the orthogonal projection of $\boldsymbol{x}$ onto the hyperplane H_i.

Exercise 5.2 Consider the following overdetermined system:

$$\begin{cases} 2x + y - 2 = 0 & (l_1), \\ 2x - y + 2 = 0 & (l_2), \\ \quad y = 0 & (l_3). \end{cases} \tag{5.14}$$

(a) Show that the least square solution to the system (5.14) uniquely exists.
(b) Give the unique least square solution to the system (5.14).
(c) Take any point $\boldsymbol{x}_0 \in \mathbb{R}^2$ define the sequence $\{\boldsymbol{x}_k\}$ by

$$\boldsymbol{x}_k := P\boldsymbol{x}_{k-1} \equiv P_3 P_2 P_1 \boldsymbol{x}_{k-1}, \ k = 1, 2, \cdots, \tag{5.15}$$

where $P_j\boldsymbol{x}$ is the orthogonal projection of $\boldsymbol{x}$ onto the line l_j $(j = 1, 2, 3)$. Calculate $\boldsymbol{x}_4$ and $\frac{1}{3}(P_1\boldsymbol{x}_4 + P_2P_1\boldsymbol{x}_4 + P_3P_2P_1\boldsymbol{x}_4)$.
(d) Which of $\boldsymbol{x}_4$ and $\frac{1}{3}(P_1\boldsymbol{x}_4 + P_2P_1\boldsymbol{x}_4 + P_3P_2P_1\boldsymbol{x}_4)$ is the closer to the unique least square solution to (5.14) calculated in (b)?
(e) Write a program using your favorite language and give a numerical simulation to compare $\boldsymbol{x}_n$ and $\frac{1}{3}(P_1\boldsymbol{x}_n + P_2P_1\boldsymbol{x}_n + P_3P_2P_1\boldsymbol{x}_n)$ for $n \geq 5$.

Exercise 5.3 Prove Proposition 5.1.

Chapter 6
Theoretical Proofs of How Approximate the Least-Square Solution is to the Real One

Abstract In this chapter, we study how approximate the least square solution is to the real solution.

In this chapter, we theoretically study how approximate the least square solution is to the real solution. Let us first review how the system (2.12) was obtained in practical applications.

In this chapter, knowledge of elementary probability and statistics is required. The readers who have not learned elementary probability and statistics can skip this chapter.

In type I problems, for example, the ones in Sects. 3.1 and 3.2, the system (2.12) is directly obtained as the mathematical model of the phenomenon or its modification. In type II problems, for example, the ones in Sects. 3.3 and 3.4, we obtain the system (1.4) as the mathematical model of the phenomenon, and the system (2.12) is obtained as a semi-equivalent system to the system (1.4).

6.1 The Case of Large Measurement Data

In this chapter, we study the case where the following assumption is satisfied.

Assumption 6.1 The solution to the system (2.10) uniquely exists.

This assumption is satisfied in many type I problems if sufficiently suitable measurements are obtained, which is neither difficult nor costly in many practical problems. For example, the solution to the problem in Sect. 3.2 is unique if we have sufficient and suitable measurements, since for the system (2.10), we have assumed that all the measurements are precise and contain no error or noise. Obtaining sufficiently suitable measurements assures Assumption 2.3, which proves the uniqueness of the solution to the system (2.10).

T. Takiguchi, *The Least Square Solution to an Overdetermined System*, SpringerBriefs in Mathematical Methods, https://doi.org/10.1007/978-981-92-0064-1_6

We also remark that Assumption 6.1 is satisfied in many type II problems, without heavy costs. For example, the solutions to the problems in Sects. 3.3 and 3.4 are unique if we have sufficiently suitable measurements; in such a case the solutions to the systems (1.2) and (2.10) are the same and unique, since for the system (1.2), we have assumed that all the measurements are precise and contain no error or noise, which yields that the semi-equivalent system (2.10) to (1.2) contains no error or noise. Assumption 2.3 is satisfied by sufficient and suitable measurements, which proves the unique solvability of the system (2.10). Therefore, the uniqueness of the solution to the system (1.2) is proved by Proposition 2.2.

In practice, however, it is impossible to obtain (2.10) or (1.2) containing no error or noise by measurements. Instead, we obtain the system (2.12) or (1.3) containing errors and noises by practical measurements. If we obtain the system (1.3), then the system (2.12) is derived as a semi-equivalent system to the system (1.3). In both cases, the least square solution to the (1.3) is unique by Assumption 2.3 and Proposition 2.6, which further yields that the real solution which is given as the solution to the system (2.10) (or (1.2)) is also uniquely determined.

Therefore, in view of practical applications, it is necessary to study how approximate the unique least square solution to the system (2.12) is to the unique solution to the system (2.10) is, which shall be studied in this section.

In many applications, we know by experience that the more data obtained, in other words, the larger the number m in (2.12) is, the more approximate to the real solution the least square solution to the system (2.10) becomes.

In this section, we give a stochastic proof to this knowledge by experience.

In the system (2.12) (or equivalently (2.13)), we divide $\tilde{\boldsymbol{s}}$ into the following two components:

$$\tilde{\boldsymbol{s}} = \boldsymbol{s} + \boldsymbol{s}^X, \tag{6.1}$$

where $\boldsymbol{s} := A\boldsymbol{x}$ and $\boldsymbol{s}_X := {}^t(s_1^X, \cdots, s_m^X)$ with s_i^X, $i = 1, \cdots, m$, being random variables whose probability density functions are supported by $[-\varepsilon_i, \varepsilon_i]$ for small $\varepsilon_i > 0$ and whose expected value $E(s_i^X)$ is 0.

We have assumed that the s_i^X are the random variables whose probability density functions are supported by $[-\varepsilon_i, \varepsilon_i]$ for small $\varepsilon_i > 0$ and whose expected value $E(s_i^X)$ is 0, which is a very natural assumption, since we can a priori estimate the observation errors, the size of the support ε_i of s_i^X each for each i, and so on.

Let us divide the system (2.13), $A\boldsymbol{x} = \tilde{\boldsymbol{s}}$ into the following two systems:

$$A\boldsymbol{x} = \boldsymbol{s}, \quad A\boldsymbol{x} = \boldsymbol{s}^X. \tag{6.2}$$

The system $A\boldsymbol{x} = \boldsymbol{s}$ in (6.2) is the same as the system (2.10).

By virtue of Proposition 2.6, the least square solution $\boldsymbol{x}^X$ to the system $A\boldsymbol{x} = \boldsymbol{s}^X$ is obtained by solving the system (2.19) having the unique solution, which yields that each component of

$$x_m^X = ({}^tAA)^{-1}\,{}^tAs^X \tag{6.3}$$

can be regarded as the arithmetical mean of m-random variables whose expectations are 0. Therefore, the (strong) law of large numbers gives that

$$\lim_{m\to\infty} x_m^X = \mathbf{0} \in \mathbb{R}^n. \tag{6.4}$$

Hence, we have proved that if $m \to \infty$, then the least square solution to the system (2.12) stochastically converges to the unique solution to the system (2.10).

Theorem 6.1 *We know by experience that the more data obtained, in other words, the larger the number m in (2.12) is, the more approximate to the real solution of (2.10), the least square solution of the system (2.12) becomes. Equation (6.4) stochastically proves that our knowledge by experience is theoretically right.*

6.2 The Case of Small Measurement Data

In the previous section, we studied how approximate the least square solution is to the real solution in the case where sufficiently many measurements are obtained. For the problems in Sects. 3.3 and 3.4, however, we know by experience that sufficient approximations of the x- and y-coordinates of the object (x, y, z) are obtained by less than 10 measurement data, for which see Proposition 4.1 in Sect. 4.3. In the problems in Sects. 3.3 and 3.4, the reconstruction of the z-coordinates being unstable, it can be corrected by the discussion in Sect. 4.4. Therefore, in the case of small measurement data, the following theorem is known by experience.

Theorem 6.2 *For the problem satisfying Property 4.1, the object can be approximately reconstructed by less than 10 measurement data, if the problem is stable in the x- and y-coordinates and even if it is unstable in the z-coordinate.*

The complete theoretical proof of Theorem 6.2 has not been given yet, for which review the discussion in Chap. 4.

Exercises
The exercises in this chapter are reviews of the context in this chapter. Therefore, the solutions are omitted.

Exercise 6.1 Review the (strong) law of large numbers and prove (6.4).

Exercise 6.2 Review the discussion in Chap. 4 and summarize what Theorem 6.2 is about.

Chapter 7
A Theoretical Study of the Algorithm for CT by G. N. Hounsfield

Abstract In this chapter, we give a theoretical justification of the idea by G. N. Hounsfield for the practical implementation of computerized tomography. He developed a device for medical CT, for which he was awarded the Nobel Prize for Physiology or Medicine. There exist a number of studies concerning Hounsfield's idea, few of which justify why Hounsfield's idea worked well at the initial stage of practical implementation of CT. We give a new stochastic approach to prove that the idea by G. N. Hounsfield is theoretically right.

As we have seen in Sect. 3.1, G. N. Hounsfield reduced the mathematical problem of CT to give the least square solution to the overdetermined system of the linear equations (3.21) (equivalently, (3.22)). In this chapter, we shall study the effect of the errors ε_i's in (3.20) in the approximate reconstruction of $x_1, x_2, \cdots, x_n$, which shall give a theoretical proof of why Hounsfield's idea worked well in the practical implementation of CT at its initial stage. The main part of this chapter has been published in the paper [22]. In this chapter, the readers are supposed to know the elements of Lebesgue integral and statistics. The readers who have not learned the elements of Lebesgue integral and statistics can skip this chapter.

7.1 Review of the Idea by G. N. Hounsfield

In this section, we review the idea by G. N. Hounsfield to derive the system (3.21) (equivalently, (3.22)) and give necessary remarks on his idea in order to prove why his idea works well. He approximated the function f of distribution of the densities restricted to the plane by a pixel function (3.18) with (3.19). Since $\inf_{(x,y)\in c_j} f(x, y) \leq x_j \leq \sup_{(x,y)\in c_j} f(x, y)$, by the definition of the Lebesgue integral, it holds that

$$\lim_{c\to 0} g_c(x, y) = f(x, y) \quad \text{in } L^1(\mathbb{R}^2). \tag{7.1}$$

T. Takiguchi, *The Least Square Solution to an Overdetermined System*, SpringerBriefs in Mathematical Methods, https://doi.org/10.1007/978-981-92-0064-1_7

In practical CT, the rank of the coefficient matrix A in (3.21) must be n. Therefore, by Theorem 2.1, we have the following theorem.

Theorem 7.1 *In the practical CT, there always exists the unique least square solution to the system (3.21) of the linear equations.*

In the rest of this chapter, we assume the following conditions:

(I) $f(x, y) \in L^1 \cap L^2(\mathbb{R}^2)$.
(II) For any line $l \subset \mathbb{R}^2$, $f|_l \in L^1(l)$ and is piecewise continuous.

Note that these conditions, (I) and (II), are satisfied by the density distribution of the human body. We also assume that

(III) $m \gg n$.

With the assumptions, (I), (II), and (III), we shall give a theoretical justification of Hounsfield's idea by applying a stochastic approach.

7.2 Effects of the Errors in the Right-Hand Sides of (3.21)

In the system (3.21), we divide the right-hand side into the following three components,

$$\boldsymbol{s} = \boldsymbol{s}^0 + \boldsymbol{s}_c^{app} + \boldsymbol{s}_c^{obs}, \tag{7.2}$$

where $\boldsymbol{s}^0 := A\boldsymbol{x}^0$ with

$$\boldsymbol{x}^0 = (x_1^0, \cdots, x_n^0), \quad x_j^0 := \frac{1}{c}\int_{C_j} f(x, y)\, dx\, dy, \tag{7.3}$$

$\boldsymbol{s}_c^{app} := ((s_c^{app})_1, \cdots, (s_c^{app})_m)$ with $(s_c^{app})_i = \int_{l_i} f(x, y)\, dl - s_i^0$ and $\boldsymbol{s}_c^{obs} = \boldsymbol{s}_X := (s_1^X, \cdots, s_m^X)$ with s_i^X, $i = 1, \cdots, m$, being random variables whose probability density functions are supported by $[-\varepsilon_i, \varepsilon_i]$ for small $\varepsilon_i > 0$ and whose expected value $E(s_i^X)$ is 0. That is, we divide ε_i's in (3.20) as

$$\varepsilon_i = (s_c^{app})_i + s_i^X. \tag{7.4}$$

By assumption (II), each component $(s_c^{app})_i$ of $\boldsymbol{s}_c^{app}$ is very small for sufficiently small $c > 0$, more precisely,

$$\lim_{c \to 0} \sum_j a_{ij} x_j^0 = \int_{l_i} f(x, y)\, dl, \tag{7.5}$$

($\lim_{c \to 0} s_i^0 = \int_{l_i} f(x, y)\, dl$) holds by the construction of the Lebesgue integral, where n tends to ∞ in (7.5).

We have assumed that the s_i^X are the random variables whose probability density functions are supported by $[-\varepsilon_i, \varepsilon_i]$ for small $\varepsilon_i > 0$ and whose expected value $E(s_i^X)$ is 0, which is a very natural assumption. In this setting, we claim that $\boldsymbol{x}^0$ is a natural theoretical solution to the system $A\boldsymbol{x}^0 = \boldsymbol{s}^0$ assigning the integral mean value of the function x_j^0 in each pixel c_j, that is, we take the pixel function as

$$g_c^0(x, y) := \sum_{j=1}^{n} x_j^0 \chi_{c_j}(x, y). \tag{7.6}$$

The least square solution $\boldsymbol{x}^{app}$ to the system $A\boldsymbol{x} = \boldsymbol{s}_\varepsilon^{app}$ gives the effect of the approximation of the function f by the pixel function (7.6), on the least square solution to the system $A\boldsymbol{x} = \boldsymbol{s}$. The least square solution $\boldsymbol{x}^{obs}$ of the system $A\boldsymbol{x} = \boldsymbol{s}_\varepsilon^{obs}$ gives the effect of the observation errors, numerical rounding errors by computers and so on. We note that the assumptions on s_i^X are natural since we can a priori estimate the observation errors, the size of the support ε_i of s_i^X each for each i, and so on. In the same way as (7.6), we define the pixel function determined by $\boldsymbol{x}^{app} = {}^t(x_1^{app}, \cdots, x_n^{app})$ and by $\boldsymbol{x}^{obs} = {}^t(x_1^{obs}, \cdots, x_n^{obs})$, respectively as

$$\begin{aligned} g_c^{app}(x, y) &:= \sum_{j=1}^{n} x_j^{app} \chi_{c_j}(x, y), \\ g_c^{obs}(x, y) &:= \sum_{j=1}^{n} x_j^{obs} \chi_{c_j}(x, y). \end{aligned} \tag{7.7}$$

By their construction, it is proved that

$$\begin{aligned} \lim_{c \to 0} g_c^0(x, y) &= f(x, y), \\ \lim_{c \to 0} g_c^{app}(x, y) &= 0, \end{aligned} \tag{7.8}$$

at almost every point (x, y). By virtue of Proposition 2.6, the least square solution $\boldsymbol{x}^{obs}$ of the system $A\boldsymbol{x} = \boldsymbol{s}_\varepsilon^{obs}$ is obtained by solving the system (2.19) having a unique solution, which yields that each component of

$$\boldsymbol{x}^{obs} = ({}^tAA)^{-1}\, {}^tA\boldsymbol{s} \tag{7.9}$$

can be regarded as the arithmetical mean of m random variables whose expectations are 0. Therefore, the (strong) law of large numbers gives that

$$\lim_{m \to \infty} g_n^{obs}(x, y) = 0, \quad a.s., \tag{7.10}$$

even for a fixed number n.

7.3 Main Theorems in This Chapter

By the arguments in the previous section, we obtain the following theorems.

Theorem 7.2 *We know by experience that even for a fixed number n, the larger the number m is, the better resolution we can obtain. Equation (7.10) stochastically proves that our knowledge by experience is theoretically right.*

Theorem 7.3 *The pixel function*

$$g_c(x, y) := \sum_{j=1}^{n} x_j \chi_{c_j}(x, y) \tag{7.11}$$

converges almost surely to the density distribution function $f(x, y)$, that is,

$$\lim_{c \to 0} g_c(x, y) = f(x, y), \quad a.s. \tag{7.12}$$

Note that by the assumption (III), $c \to 0$ yields $m \gg n \to \infty$. Remember also that

$$g_c(x, y) = g_c^0(x, y) + g_c^{app}(x, y) + g_c^{obs}(x, y). \tag{7.13}$$

Theorems 7.2 and 7.3 stochastically prove that the idea by Hounsfield is theoretically right, in the sense that his solution approximates the unique least square solution to (3.21) if the number n of the pixels is sufficiently large. We have a large enough number $m \gg n$ of the X-ray data and the errors are sufficiently small, which can be naturally assumed in usual practical applications.

Exercises

The exercises in this chapter are reviews of the context in this chapter. Therefore, the solutions are omitted.

Exercise 7.1 Answer the following problems:

(a) Review the construction of the Lebesgue integral and prove (7.1).
(b) Review the (strong) law of large numbers and prove (7.10).

Exercise 7.2 Answer the following problems:

(a) Summarize the proof of Theorem 7.2.
(b) Summarize the proof of Theorem 7.3.

Chapter 8
Ultrasonic Imaging (USI)

Abstract In this chapter, we introduce a challenging problem to establish a new non-invasive and/or non-destructive inspection technique by application of ultrasound.

Let us introduce a challenging problem to establish a new non-invasive and/or non-destructive inspection technique by application of ultrasound, *ultrasonic imaging* (USI).

As non-invasive and/or non-destructive inspection techniques, X-ray computerized tomography and magnetic resonance imaging (MRI) are well known and frequently applied. X-ray CT and MRI are very superior techniques in non-invasive and non-destructive inspections, since their measurement errors are very small and very sharp reconstruction techniques are developed. In the measurements of USI, larger errors are included than CT and MRI, and a reconstruction technique as sharp as CT and/or MRI has not been developed. Therefore, it is almost impossible that USI gives the reconstruction images better than or equal to CT or MRI. The X-ray CT and the MRI, however, have shortcomings such as: they require very expensive protection facilities, they are very expensive to run, and X-rays and huge electromagnetic fields have bad effects on the human bodies and the environment. USI gives satisfactory reconstruction images, not good as CT and MRI, with very cheap running cost, no protection facilities and no negative effect on human bodies or the environment. Therefore, the development of USI is important to provide suitable medical imaging techniques to the areas where the introduction of CT or MRI is difficult for various reasons or conflict areas where medical facilities and devices are in short supply.

The term "tomography" is often used for imaging techniques, such as, acoustic tomography, electrical impedance tomography, optical tomography, photoacoustic tomography and so on. As we have seen in Sect. 3.1, the meaning of the term "tomography" is "to draw a picture of the section of the object by a computer". In X-ray CT, the reconstructed image is the distribution of the densities of the human body *restricted to the plane*, therefore it is a *tomograph*. In USI, we do not give an image

T. Takiguchi, *The Least Square Solution to an Overdetermined System*, SpringerBriefs in Mathematical Methods, https://doi.org/10.1007/978-981-92-0064-1_8

of the section, but try to reconstruct the whole three-dimensional image. Therefore, we would call our technique *ultrasonic imaging*, *not ultrasonic tomography*. Of course, if we obtain the whole 3D image of the object, then we can take its 2D section images. In USI, our main purpose is to obtain the whole 3D image of the object, which is better than the main purpose of CT.

The essential parts of this chapter have been published in the papers [11, 20, 22].

8.1 Mathematical Problem of USI

USI is a kind of an acoustic imaging. Many people call acoustic imaging *acoustic tomography*, however, we call it *acoustic imaging* for the reason mentioned above. In USI, we apply ultrasound whose frequency differs in accordance with the object for inspection. If the object is concrete structures, we apply ultrasound of a frequency about 50 kHz, while ultrasound of a frequency about 2 MHz is applied for the human body. Though we are able to study the same inverse problems in acoustic imaging, in this chapter, we study other problems than acoustic imaging, which are very special to our USI problem.

Property 8.1 Let $\Omega \subset \mathbb{R}^3$ be a domain where the object is situated, and $f(x)$ be the propagation speed of the ultrasound at the point $x \in \Omega$. For $\alpha, \beta \in \partial\Omega$, we denote by $\gamma_{\alpha,\beta}$ a route from α to β contained in Ω. The primary wave of the ultrasound which travels from the point $\alpha \in \partial\Omega$ to the point $\beta \in \partial\Omega$ takes the route where the travel time is given by

$$\min_{\gamma_{\alpha,\beta}} \int_{\gamma_{\alpha,\beta}} 1/f(x)\, d\gamma \tag{8.1}$$

This route is called "the fastest route".

For this property, cf. [11]. By virtue of this property, the mathematical problem for USI is given as:

Problem 8.1 (Problem to Develop USI) Let $\Omega \subset \mathbb{R}^3$ be a domain and $f(x)$ be the propagation speed of the ultrasonic wave at the point $x \in \Omega$. For $\alpha, \beta \in \partial\Omega$, we denote by $\gamma_{\alpha,\beta}$ a route from α to β contained in Ω. In this case, reconstruct $f(x)$ $(x \in \Omega)$ out of the data

$$\min_{\gamma_{\alpha,\beta}} \int_{\gamma_{\alpha,\beta}} 1/f(x)\, d\gamma, \tag{8.2}$$

for $\forall\alpha, \beta \in \partial\Omega$.

Remark 8.1 In many imaging techniques with ultrasound, measurements of many types of waves are applied, for example, the first arriving waves, the reflected waves, the refracted waves and so on. In Problem 8.1, it is new that we focus only on the first arriving waves. In this way, the data for the reconstruction of the object are much smaller than in other imaging techniques with ultrasound. However, the mathematical model for analysis becomes much simpler, which is the biggest advantage to Problem 8.1.

Problem 8.1 was originally posed in [11]. The mathematical solution to Problem 8.1 being very difficult and open, we would like to pose a practical problem related to Problem 8.1 and solve it. In the paper [11], the authors posed and solved how to non-destructively inspect the concrete cover for reinforcement in reinforced concrete structures.

If the object is small or close to homogeneous, then we can apply the same algorithm with X-ray CT, that is, we assume that the ultrasound rectilinearly propagates and apply the same algorithm with X-ray CT to obtain a rough sketch of the distribution of the reciprocal $1/f(x)$ of the velocity. This idea is successful in medical imaging where the object is very small (cf. [1]). In the next section, we shall extend the this idea for further development of USI.

8.2 A Numerical Approach to Problem 8.1

In this section, we propose a numerical approach to Problem 8.1 based on Property 8.1 and Hounsfield's idea.

Method 8.1 For a numerical approach to Problem 8.1, we propose the following iterative numerical method:

(1) Divide the object into the direct sum of congruent cube boxels.
(2) Assuming that the ultrasound rectilinearly propagates, apply G. N. Hounsfield's idea to obtain the set g_1 of the boxel functions as the least square solution to (2.10), which approximates $1/f(x)$, the reciprocal of the sonic velocity $f(x)$ in each boxel. In this procedure, the elements a_{ij} of the matrix in (2.10) are given by the length of intersection between the linear (assumed) orbit of the i-th ultrasound and j-th boxel and s_i is the travel time of the i-th ultrasound.
(3) Take the boxel function g_1 obtained in the procedure (2) as the first image. By the first image, we can calculate the new orbit of the ultrasound. Then we change the elements a_{ij} of the matrix by the length of intersection between the new orbit of the i-th ultrasound and j-th boxel. Giving the least square solution g_2 to (2.10) with new matrix elements gives better approximation.
(4) Take the set g_2 of the boxel functions as the second image and repeat the same procedure as in (3) and obtain a better approximation boxel function g_3.
(5) Repeat the procedure (4) by changing the k-th image of the g_k of boxel functions to the new obtained one g_{k+1}, until $\sup |g_{k+1} - g_k|$ becomes sufficiently small.
(6) When $\sup |g_{k+1}-g_k|$ becomes sufficiently small then the set g_{k+1} is a numerical solution boxel function to Problem 8.1.

Let us explain this method.

In the first step (Method 8.1 (1), (2)), we assume that the ultrasound rectilinearly propagates and apply the same algorithm as G. N. Hounsfield to obtain the boxel function g_1, which gives the first image for our iterative algorithm. In this step, we assume that the ultrasound rectilinearly propagates in the whole structure as well as in each boxel. In Fig. 8.1, we give a two-dimensional picture for simplicity.

In the second step (Method 8.1 (3)), having known the distribution functions g_1 of the reciprocal of the sonic velocity, we can calculate the fastest route between the two boundary points. In this case, the route where the ultrasound propagates is not a line, in general, but it is assumed to rectilinearly propagate in each boxel (see Fig. 8.2). Therefore, the fastest route becomes a polygonal line and each component a_{ij} in the overdetermined system (2.10) (or (2.11)) is the intersection of the fastest route of the i-th ultrasound and the j-th boxel c_j. By changing the coefficients a_{ij} in the overdetermined system (2.10) (or (2.11)), we obtain a better approximation function g_2 of $1/f(x)$ than g_1, as the least square solution to the system (2.10) (or (2.11)).

In the third step (Method 8.1 (4)), we calculate the new fastest routes based on g_2 and repeat the same procedure as the second step, by replacing $g_1 \to g_2$.

In the final step (Method 8.1 (5), (6)), iteration of the same procedure by replacing $g_k \to g_{k+1}$ until $\sup |g_{k+1} - g_k|$ is sufficiently small gives that g_{k+1} is a sufficiently approximate boxel function of $1/f(x)$.

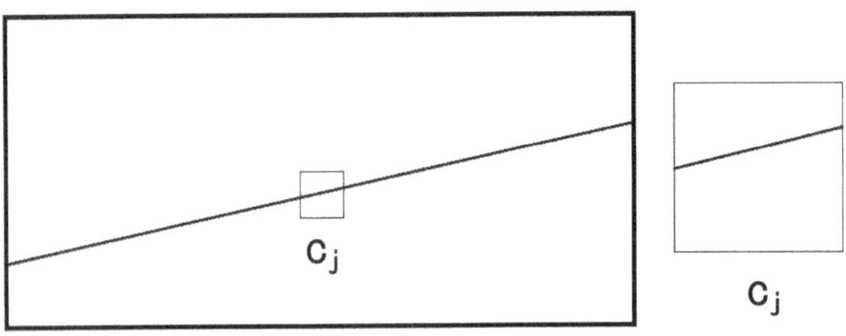

Fig. 8.1 First step of the algorithm

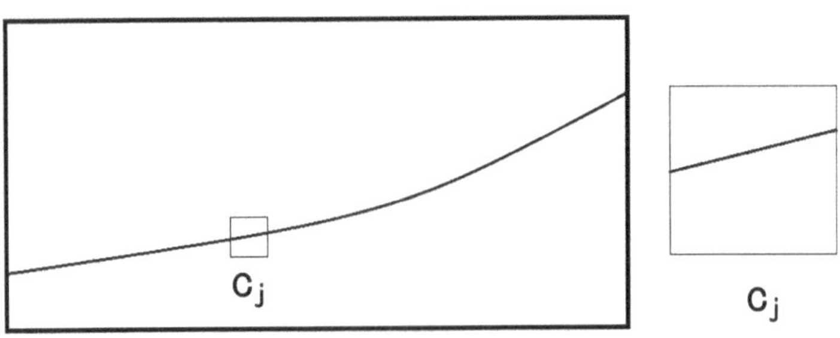

Fig. 8.2 Second step of the algorithm

Therefore, in order to practicalize Method 8.1, it is important problem to solve the following problem for the second step.

Problem 8.2 Assume that the distribution $f(x)$ of the velocity, $x \in \Omega \subset \mathbb{R}^3$ is known. For any $\alpha, \beta \in \partial\Omega$, determine the fastest route γ from α to β though Ω, that is,

$$\min_{\gamma_{\alpha,\beta}} \int_{\gamma_{\alpha,\beta}} 1/f(x)\, d\gamma = \int_{\gamma} 1/f(x)\, d\gamma. \tag{8.3}$$

This problem is challenging and interesting. The complete solution to Problem 8.2 seems to be difficult; however, it is sufficient to solve this problem by assuming that the distribution $f(x)$ of the velocity is a boxel function. Furthermore, giving an algorithm to find the fastest route for a boxel function $f(x)$ is also practically sufficient.

It is interesting to give a numerical approach to Problem 8.2 by application of AI (artificial intelligence) and/or machine learning.

Exercises

The exercises in this chapter are open problems and new research tasks. Therefore, the solutions are omitted.

Exercise 8.1 Give a solution to Problem 8.2.

Exercise 8.2 Give a numerical solution to Problem 8.2 by application of AI and/or machine learning and apply the solution to implement Method 8.1 numerically.

Appendix A
Review of Elementary Linear Algebra

In Chap. 2, we applied matrix theory to analyze overdetermined systems of linear equations. In Appendix, let us review elementary linear algebra, especially on matrix theory and on the analysis of underdetermined linear systems.

A.1 Linear Systems and Elementary Matrices

Let us consider the following linear system:

$$\begin{cases} 2x - y - z = 3, \\ x + y - 2z = 0, \\ x - z = 1. \end{cases} \tag{A.1}$$

Equation (A.1) is represented as

$$\begin{pmatrix} 2 & -1 & -1 \\ 1 & 1 & -2 \\ 1 & 0 & -1 \end{pmatrix} \begin{pmatrix} x \\ y \\ z \end{pmatrix} = \begin{pmatrix} 3 \\ 0 \\ 1 \end{pmatrix}. \tag{A.2}$$

The matrix in the left-hand side of (A.2) is called *the coefficient matrix* of the linear system (A.1). The following matrix

$$\begin{pmatrix} 2 & -1 & -1 & : & 3 \\ 1 & 1 & -2 & : & 0 \\ 1 & 0 & -1 & : & 1 \end{pmatrix} \tag{A.3}$$

consisting of the coefficient matrix and the row of the right-hand side of (A.2), is called *the augmented coefficient matrix* of the system (A.1).

T. Takiguchi, *The Least Square Solution to an Overdetermined System*, SpringerBriefs in Mathematical Methods, https://doi.org/10.1007/978-981-92-0061-1

In order to solve the system (A.1), we make use of the following three operations:

(S1) Multiply both sides of one equation by a constant $t \neq 0$.
(S2) Interchange two equations.
(S3) Add the both sides of one equation multiplied by a constant $t \neq 0$ to the counterparts of another.

These three operations are interpreted in the following operations in terms of the augmented coefficient matrix:

Definition A.1 (The Row (or Left) Elementary Operations of the Matrices) For matrices, the following operations are called *the row (or the left) elementary operations of the matrices* .

(L1) Multiply some row by a constant $c \neq 0$.
(L2) Interchange two rows.
(L3) Add a row multiplied by a constant $c \neq 0$ to another.

If we can transform the coefficient matrix of the system of linear equations with unknowns x, y, z to the unit matrix by the row elementary operations, or equivalently, if we can transform the augmented coefficient matrix by the row elementary operations to the following form,

$$\begin{pmatrix} 1\,0\,0 : \alpha \\ 0\,1\,0 : \beta \\ 0\,0\,1 : \gamma \end{pmatrix}, \tag{A.4}$$

then the system (A.1) is solved by $(x, y, z) = (\alpha, \beta, \gamma)$.

Having this in mind, let us transform the augmented coefficient matrix (A.3) by the row elementary operations:

$$\begin{pmatrix} 2 & -1 & -1 & : 3 \\ 1 & 1 & -2 & : 0 \\ 1 & 0 & -1 & : 1 \end{pmatrix} \xrightarrow[(a)]{(r1)\leftrightarrow(r2)} \begin{pmatrix} 1 & 1 & -2 & : 0 \\ 2 & -1 & -1 & : 3 \\ 1 & 0 & -1 & : 1 \end{pmatrix} \xrightarrow[(b)]{(r2)-2(r1)} \begin{pmatrix} 1 & 1 & -2 & : 0 \\ 0 & -3 & -3 & : 3 \\ 1 & 0 & -1 & : 1 \end{pmatrix}$$

$$\xrightarrow[(c)]{(r3)-(r1)} \begin{pmatrix} 1 & 1 & -2 & : 0 \\ 0 & -3 & 3 & : 3 \\ 0 & -1 & 1 & : 1 \end{pmatrix} \xrightarrow[(d)]{-\frac{1}{3}(r2)} \begin{pmatrix} 1 & 1 & -2 & : 0 \\ 0 & 1 & -1 & : -1 \\ 0 & -1 & 1 & : 1 \end{pmatrix}$$

$$\xrightarrow[(e)]{(r1)-(r2)} \begin{pmatrix} 1 & 0 & -1 & : 1 \\ 0 & 1 & -1 & : -1 \\ 0 & -1 & 1 & : 1 \end{pmatrix} \xrightarrow[(f)]{(r3)+(r2)} \begin{pmatrix} 1 & 0 & -1 & : 1 \\ 0 & 1 & -1 & : -1 \\ 0 & 0 & 0 & : 0 \end{pmatrix},$$

where (rj) stands for the j-th row for $j = 1, 2, 3$. If necessary we denote the j-th column by (cj).

Let us explain each operation:

(a) Interchange $(r1)$ and $(r2)$ in order for the $(1, 1)$ component to be 1.
(b) Subtract $2(r1)$ from $(r2)$ in order for the $(2, 1)$ component to be 0.
(c) Subtract $(r1)$ from $(r3)$ in order for the $(3, 1)$ component to be 0, which has the first column the same as the unit matrix.
(d) Multiply $(r2)$ by $-\frac{1}{3}$ in order for the $(2, 2)$ component to be 1.
(e) Subtract $(r2)$ from $(r1)$ in order for the $(1, 2)$ component to be 0.
(f) Add $(r2)$ to $(r3)$ in order for the $(3, 2)$ component to be 0, which has the first and the second column the same as the unit matrix.

We cannot go further since we cannot make the $(3, 3)$ component 1 without changing the first and the second columns. By the transformed augmented coefficient matrix, we obtain the system

$$\begin{pmatrix} 1 & 0 & -1 & : & 1 \\ 0 & 1 & -1 & : & -1 \\ 0 & 0 & 0 & : & 0 \end{pmatrix} \Leftrightarrow \begin{cases} x - z = 1, \\ y - z = -1, \\ 0 = 0. \end{cases} \tag{A.5}$$

In view of (A.4), the first row in the transformed augmented coefficient matrix determines the unknown x, the second row y and the third z. The third row or equation in (A.5) determines the unknown z; however, $0 = 0$ holds for any z. Hence, z can be taken arbitrarily. Therefore, the system (A.1) is solved by

$$\begin{pmatrix} x \\ y \\ z \end{pmatrix} = \begin{pmatrix} t + 1 \\ t - 1 \\ t \end{pmatrix}, \tag{A.6}$$

where t is an arbitrary constant.

In general, let us consider the system

$$\begin{cases} a_{11}x_1 + a_{12}x_2 + \cdots + a_{1n}x_n = s_1, \\ a_{21}x_1 + a_{22}x_2 + \cdots + a_{2n}x_n = s_2, \\ \cdots\cdots\cdots \\ a_{m1}x_1 + a_{m2}x_2 + \cdots + a_{mn}x_n = s_m, \end{cases} \tag{A.7}$$

or equivalently

$$A\boldsymbol{x} = \boldsymbol{s} \tag{A.8}$$

and its augmented coefficient matrix $(A : \boldsymbol{s})$.

Definition A.2 (Elementary Matrices) The following matrices $P_m(i; c)$, $Q_m(i, j)$, $R_m(i, j; c)$ are called *the elementary matrices*:

$$P_m(i;c) = \begin{pmatrix} 1 & & & & & & \\ & \ddots & & & & & \\ & & 1 & & & & \\ & & & c & & & \\ & & & & 1 & & \\ & & & & & \ddots & \\ & & & & & & 1 \end{pmatrix}, \tag{A.9}$$

$$Q_m(i,j) = \begin{pmatrix} 1 & & & & & & \\ & \ddots & & & & & \\ & & 0 & & 1 & & \\ & & & \ddots & & & \\ & & 1 & & 0 & & \\ & & & & & \ddots & \\ & & & & & & 1 \end{pmatrix}, \tag{A.10}$$

$$R_m(i,j;c) = \begin{pmatrix} 1 & & & & & & \\ & \ddots & & & & & \\ & & 1 & & c & & \\ & & & \ddots & & & \\ & & & & 1 & & \\ & & & & & \ddots & \\ & & & & & & 1 \end{pmatrix}. \tag{A.11}$$

$P_m(i;c), Q_m(i,j), R_m(i,j;c)$ are $m \times m$ matrices and they are represented coefficient-wise such that

$$(P_m(i;c))_{k,l} = \begin{cases} 1 \ (k = l \neq i), \\ c \ (k = l = i), \\ 0 \quad (k \neq l), \end{cases}$$

$$(Q_m(i,j))_{k,l} = \begin{cases} 1 \quad (k = l \neq i \text{ or } j), \\ 1 \ ((k,l) = (i,j) \text{ or } (j,i)), \\ 0 \quad (other), \end{cases} \tag{A.12}$$

$$(R_m(i,j;c))_{k,l} = \begin{cases} 1 \quad (k = l), \\ c \ (k = i,\ l = j), \\ 0 \quad (other), \end{cases}$$

where $(A)_{k,l}$ is the (k,l) element of the matrix A and $c \neq 0$ is a constant.

The operation (L1) is realized by multiplying the matrix $P_m(i; c)$ from the left, the operation (L2) is realized by multiplying the matrix $Q_m(i, j)$ from the left, and the operation (L3) is realized by multiplying the matrix $R_m(i, j; c)$ from the left.

A.2 The Elementary Operations and the Rank of the Matrix

Definition A.3 (The Elementary Operations of the Matrices) For matrices, the following operations are called *the elementary operations of the matrices* .

(L1) Multiply the i-th row by a constant $c \neq 0$.
(L2) Interchange the i-th and the j-th rows.
(L3) Add the j-th row multiplied by a constant $c \neq 0$ to the i-th row.
(R1) Multiply the i-th column by a constant $c \neq 0$.
(R2) Interchange i-th and the j-th columns.
(R3) Add i-th column multiplied by a constant $c \neq 0$ to the j-th column.

For an $m \times n$ matrix A:

- The operation (L1) is realized by multiplying the matrix $P_m(i; c)$ from the left $(P_m(i; c)A)$.
- The operation (L2) is realized by multiplying the matrix $Q_m(i, j)$ from the left $(Q_m(i, j)A)$.
- The operation (L3) is realized by multiplying the matrix $R_m(i, j; c)$ from the left $(R_m(i, j; c)A)$.
- The operation (R1) is realized by multiplying the matrix $P_n(i; c)$ from the right $(AP_n(i; c))$.
- The operation (R2) is realized by multiplying the matrix $Q_n(i, j)$ from the right $(AQ_n(i, j))$.
- The operation (R3) is realized by multiplying the matrix $R_n(i, j; c)$ from the right $(AR_n(i, j; c))$.

Proposition A.1 *The elementary matrices are regular and their inverse matrices are given by*

$$P_m(i; c)^{-1} = P_m(i; \frac{1}{c}), \quad Q_m(i, j)^{-1} = Q_m(i, j), \quad R_m(i, j; c)^{-1} = R_m(i, j; -c). \tag{A.13}$$

Proposition A.2 *Any $m \times n$ matrix is transformed to the following canonical form $F_{m,n}(r)$ uniquely by the elementary operations.*

$$F_{m,n}(r) = \begin{pmatrix} I_r & O_{r,n-r} \\ O_{m-r,r} & O_{m-r,n-r} \end{pmatrix}, \tag{A.14}$$

where I_r is the $r \times r$ unit matrix and $O_{k,l}$ is the $k \times l$ zero matrix.

Proof It is not difficult to show that any $m \times n$ matrix is transformed to the following form:

$$\begin{pmatrix} I_r & B_{r,n-r} \\ O_{m-r,r} & O_{m-r,n-r} \end{pmatrix}, \tag{A.15}$$

where $B_{r,n-r}$ is some $r \times (n-r)$ matrix, by the elementary operations (L1), (L2), (L3), and (R2). Therefore, the canonical form (A.14) is obtained by the elementary operation (R3) necessary times to (A.15).

Let us show the uniqueness of the canonical form. Assume that an $m \times n$ matrix A is transformed to two canonical forms

$$S_i \cdots S_1 A T_1 \cdots T_j = F_{m,n}(r), \quad U_k \cdots U_1 A V_1 \cdots V_l = F_{m,n}(s), \tag{A.16}$$

where $S_1, \cdots, S_i, T_1, \cdots, T_j, U_1, \cdots, U_k, V_1, \cdots, V_l$ are elementary matrices and $r \leq s$. Since elementary matrices are regular, $S = S_i \times \cdots \times S_1, U = U_k \times \cdots \times U_1$ are $m \times m$ regular matrices and $T = T_1 \times \cdots \times T_j$, $V = V_1 \times \cdots \times V_l$ are $n \times n$ regular matrices. By (A.16), we obtain

$$F_{m,n}(s) = US^{-1}F_{m,n}(r)T^{-1}V. \tag{A.17}$$

In (A.17), $P = US^{-1}$ is an $m \times m$ regular matrix and $Q = T^{-1}V$ is an $n \times n$ regular matrix. Let us write

$$P = \begin{pmatrix} P_{11} & P_{12} \\ P_{21} & P_{22} \end{pmatrix}, \quad Q = \begin{pmatrix} Q_{11} & Q_{12} \\ Q_{21} & Q_{22} \end{pmatrix}, \tag{A.18}$$

where P_{11} and Q_{11} are $r \times r$ matrices, P_{12} is $r \times (m-r)$, P_{21} is $(m-r) \times r$, P_{22} is $(m-r) \times (m-r)$, Q_{12} is $r \times (n-r)$, Q_{21} is $(n-r) \times r$ and Q_{22} is $(n-r) \times (n-r)$.

By virtue of (A.17) and (A.18), it holds that

$$\begin{pmatrix} I_s & O_{s,n-s} \\ O_{m-s,s} & O_{m-s,n-s} \end{pmatrix} = \begin{pmatrix} P_{11} & P_{12} \\ P_{21} & P_{22} \end{pmatrix} \begin{pmatrix} I_r & O_{r,n-r} \\ O_{m-r,r} & O_{m-r,n-r} \end{pmatrix} \begin{pmatrix} Q_{11} & Q_{12} \\ Q_{21} & Q_{22} \end{pmatrix}. \tag{A.19}$$

The right-hand side of (A.19) is calculated as

$$\begin{aligned} &\begin{pmatrix} P_{11} & P_{12} \\ P_{21} & P_{22} \end{pmatrix} \begin{pmatrix} I_r & O_{r,n-r} \\ O_{m-r,r} & O_{m-r,n-r} \end{pmatrix} \begin{pmatrix} Q_{11} & Q_{12} \\ Q_{21} & Q_{22} \end{pmatrix} \\ &\quad = \begin{pmatrix} P_{11} & O_{r,n-r} \\ P_{21} & O_{m-r,n-r} \end{pmatrix} \begin{pmatrix} Q_{11} & Q_{12} \\ Q_{21} & Q_{22} \end{pmatrix} = \begin{pmatrix} P_{11}Q_{11} & P_{11}Q_{12} \\ P_{21}Q_{11} & P_{21}Q_{12} \end{pmatrix}. \end{aligned} \tag{A.20}$$

Therefore we obtain

$$\begin{pmatrix} I_s & O_{s,n-s} \\ O_{m-s,s} & O_{m-s,n-s} \end{pmatrix} = \begin{pmatrix} P_{11}Q_{11} & P_{11}Q_{12} \\ P_{21}Q_{11} & P_{21}Q_{12} \end{pmatrix}. \tag{A.21}$$

Equation (A.21) and the fact $r \leq s$ yield that

$$P_{11}Q_{11} = I_r, \quad P_{11}Q_{12} = O_{r,n-r}, \quad P_{21}Q_{11} = O_{m-r,r}. \tag{A.22}$$

By the fact $P_{11}Q_{11} = I_r$, P_{11} and Q_{11} are regular. Multiplying both hands of $P_{11}Q_{12} = O_{r,n-r}$ by P_{11}^{-1} from the left, and multiplying both hands of $P_{21}Q_{11} = O_{m-r,r}$ by Q_{11}^{-1} the from the right we obtain

$$Q_{12} = O_{r,n-r}, \quad P_{21} = O_{m-r,r}, \tag{A.23}$$

which shows

$$P_{21}Q_{12} = O_{m-r,r}O_{r,n-r} = O_{m-r,n-r}. \tag{A.24}$$

By (A.19), (A.20), (A.21), (A.22), (A.23), and (A.24), we obtain

$$F_{m,n}(s) = \begin{pmatrix} I_s & O_{s,n-s} \\ O_{m-s,s} & O_{m-s,n-s} \end{pmatrix} = \begin{pmatrix} I_r & O_{r,n-r} \\ O_{m-r,r} & O_{m-r,n-r} \end{pmatrix} = F_{m,n}(r), \tag{A.25}$$

which proves $s = r$ and the uniqueness of the canonical form. □

Definition A.4 (The Rank of the Matrix) Any $m \times n$ matrix A is transformed to the unique canonical form $F_{m,n}(r)$ defined in (A.14) by the elementary operations. In this case, the rank of the matrix A is defined to be r, which is denoted by

$$\text{rank}(A) = r. \tag{A.26}$$

A.3 The Solutions to a Linear System

Let us go back to the linear system (A.7) (or (A.8)) whose augmented coefficient matrix is $(A : \mathbf{s})$.

Proposition A.3 (Solutions to the Linear System (A.7)) *The coefficient matrix A of the linear system (A.7) is transformed to the following form of (A.15),*

$$\begin{pmatrix} I_r & B_{r,n-r} \\ O_{m-r,r} & O_{m-r,n-r} \end{pmatrix}, \tag{A.27}$$

by the elementary operations (L1), (L2), (L3), and (R2). In this case, rank(A) = r. Similarly, the augmented coefficient matrix is $(A : \mathbf{s})$ is transformed to the following form,

$$\begin{pmatrix} I_r & B_{r,n-r} & : \mathbf{s}' \\ O_{m-r,r} & O_{m-r,n-r} & : \mathbf{s}'' \end{pmatrix}, \tag{A.28}$$

by the elementary operations (L1), (L2), (L3) and the interchange of two columns other than the last one, where $\mathbf{s}' \in \mathbb{R}^r$ *and* $\mathbf{s}'' \in \mathbb{R}^{m-r}$. *If* $\mathbf{s}'' \neq \mathbf{0}$ *(in this case,* $\mathrm{rank}(A : \mathbf{s}) > r$*), then the system (A.7) has no solution. If* $\mathbf{s}'' = \mathbf{0}$ *(in this case,* $\mathrm{rank}(A : \mathbf{s}) = r$*), then the augmented coefficient matrix* $(A : \mathbf{s})$ *is transformed to the following form,*

$$\begin{pmatrix} I_r & B_{r,n-r} & : \mathbf{s}' \\ O_{m-r,r} & O_{m-r,n-r} & : \mathbf{0} \end{pmatrix} = \begin{pmatrix} 1 & & & b_{1(r+1)} & \cdots & b_{1n} & : & s_1' \\ & \ddots & & \vdots & & \vdots & \vdots & \vdots \\ & & 1 & b_{r(r+1)} & \cdots & b_{rn} & : & s_r' \\ 0 & \cdots & 0 & 0 & \cdots & 0 & : & 0 \\ \vdots & \ddots & \vdots & \vdots & \ddots & \vdots & : & \vdots \\ 0 & \cdots & 0 & 0 & \cdots & 0 & : & 0 \end{pmatrix}, \tag{A.29}$$

by the elementary operations (L1), (L2), (L3) and the interchange of two columns other than the last one.

In this case, the solution to the system (A.7) is

$$\begin{pmatrix} x_1 \\ \vdots \\ x_r \\ x_{r+1} \\ \vdots \\ x_n \end{pmatrix} = \begin{pmatrix} s_1' - b_{1(r+1)}\alpha_{r+1} \cdots -b_{1n}\alpha_n \\ \vdots \\ s_r' - b_{r(r+1)}\alpha_{r+1} \cdots -b_{rn}\alpha_n \\ \alpha_{r+1} \\ \vdots \\ \alpha_n \end{pmatrix}, \tag{A.30}$$

where $\alpha_{r+1}, \cdots, \alpha_n$ *are arbitrary constants.*

Remark A.1 In Proposition A.3, note that the application of the interchange of two columns interchanges the variables. For example, if we interchange (ri) and (rj) to transform the augmented coefficient matrix to (A.29) then x_i and x_j in (A.30) are interchanged.

Let us study the structure of the solutions (A.30) to the system (A.7). Consider the following system of homogeneous linear equations,

$$\begin{cases} a_{11}x_1 + a_{12}x_2 + \cdots + a_{1n}x_n &= 0, \\ a_{21}x_1 + a_{22}x_2 + \cdots + a_{2n}x_n &= 0, \\ \cdots\cdots\cdots \\ a_{m1}x_1 + a_{m2}x_2 + \cdots + a_{mn}x_n &= 0, \end{cases} \tag{A.31}$$

or equivalently

$$A\boldsymbol{x} = \mathbf{0}, \tag{A.32}$$

whose augmented coefficient matrix is $(A : \mathbf{0})$.

Definition A.5 (Linear Subspaces) A subset $V \subset \mathbb{R}^n$ is a linear subspace of $\mathbb{R}^n$ if and only if

$$\boldsymbol{x} + \boldsymbol{y} \in V \quad \text{and} \quad \alpha \boldsymbol{x} \in V, \tag{A.33}$$

for any $\boldsymbol{x}, \boldsymbol{y} \in V$ and for any $\alpha \in \mathbb{R}$.

Proposition A.4 (Structure of the Solutions to the System (A.31)) *If the rank of the coefficient matrix A is r, then the solutions to the system (A.31) become an $(n - r)$ dimensional linear subspace of $\mathbb{R}^n$.*

This proposition is proved by the fact that the solution to the system (A.31) is given by

$$\begin{pmatrix} x_1 \\ \vdots \\ x_r \\ x_{r+1} \\ \vdots \\ x_n \end{pmatrix} = \begin{pmatrix} -b_{1(r+1)}\alpha_{r+1} \cdots -b_{1n}\alpha_n \\ \vdots \\ -b_{r(r+1)}\alpha_{r+1} \cdots -b_{rn}\alpha_n \\ \alpha_{r+1} \\ \vdots \\ \alpha_n \end{pmatrix}, \tag{A.34}$$

where $\alpha_{r+1}, \cdots, \alpha_n$ are arbitrary constants.

Remark A.2 We give the same remark as Remark A.1. In Proposition A.4, note that the application of the interchange of two columns interchanges the variables.

Definition A.6 (The Affine Space) If W is obtained by the parallel translation of the linear subspace $V \subset \mathbb{R}^n$ by a vector $\boldsymbol{w} \in \mathbb{R}^n$, then,

$$W = \boldsymbol{w} + V \tag{A.35}$$

holds. In this case, W is called *an affine space* and V is called *the linear subspace associated with W* .

Proposition A.5 (Structure of the Solutions to the System (A.7)) *Take one solution $\boldsymbol{x}_0$ to the system (A.7) and the following holds:*

$$\{\textit{any solution to (A.7)}\} = \{\boldsymbol{x}_0 + (\textit{any solution to (A.31)})\}. \tag{A.36}$$

Therefore, the solutions to the system (A.7) are an affine space and the linear subspace of the solutions to the system (A.31) is its associated linear subspace.

A.4 Linear System and Square Matrix

In this section, let us consider the following linear system,

$$\begin{cases} a_{11}x_1 + a_{12}x_2 + \cdots + a_{1n}x_n = s_1, \\ a_{21}x_1 + a_{22}x_2 + \cdots + a_{2n}x_n = s_2, \\ \cdots\cdots\cdots \\ a_{n1}x_1 + a_{n2}x_2 + \cdots + a_{nn}x_n = s_n, \end{cases} \tag{A.37}$$

or equivalently

$$A\boldsymbol{x} = \boldsymbol{s}, \tag{A.38}$$

where the coefficient matrix A is a square $n \times n$ matrix.

Proposition A.6 *If the rank of an $n \times n$ matrix A is n, then A can be transformed to the unit matrix I_n only by the elementary operations (L1), (L2), and (L3).*

By virtue of this proposition, we have:

Proposition A.7 *For an $n \times n$ matrix A, the following conditions are equivalent:*

(i) *$rank(A) = n$.*
(ii) *The matrix A is regular.*

For the proofs of Propositions A.6 and A.7, try Exercise A.4 below.

By Propositions A.6 and A.7, we obtain the following.

Proposition A.8 *If the coefficient matrix A of the system (A.37)(or (A.38)) is regular, then the system (A.37) (or (A.38)) is uniquely solved by*

$$\boldsymbol{x} = A^{-1}\boldsymbol{s}. \tag{A.39}$$

Exercises

Exercise A.1

(a) Prove Proposition A.1.
(b) Complete the proof of Proposition A.2.

Exercise A.2 Prove Proposition A.4.

Exercise A.3 Prove Proposition A.5.

Exercise A.4

(a) Prove Proposition A.6.
(b) Prove Proposition A.7.

Solutions to the Exercises and with Comments

Chapter 1

1.1 (b) and (c). (b) is an inverse problem to inversely solve the flow of time and the cause-and-effect relation. (c) is an inverse problem to inversely solve the cause-and-effect relation.

1.2 There are a number of examples of inverse problems.

Chapter 2

2.1 This is an simple exercise of linear algebra.

$$\text{Let } \boldsymbol{x} = \begin{pmatrix} x_1 \\ x_2 \\ \vdots \\ x_n \end{pmatrix}, \boldsymbol{y} = \begin{pmatrix} y_1 \\ y_2 \\ \vdots \\ y_m \end{pmatrix} \text{ and } B = \begin{pmatrix} b_{11} & b_{12} & \dots & b_{1n} \\ b_{21} & b_{22} & \cdots & b_{2n} \\ \vdots & \vdots & \ddots & \vdots \\ b_{m1} & b_{m2} & \cdots & b_{mn} \end{pmatrix}.$$

(i) Calculate the vector $B\boldsymbol{x} \in \mathbb{R}^m$.
(ii) Calculate the inner product $\langle B\boldsymbol{x}, \boldsymbol{y}\rangle_{\mathbb{R}^m}$ in $\mathbb{R}^m$.
(iii) Calculate the vector ${}^tB\boldsymbol{y} \in \mathbb{R}^n$.
(iv) Calculate the inner product $\langle \boldsymbol{x}, {}^tB\boldsymbol{y}\rangle_{\mathbb{R}^n}$ in $\mathbb{R}^n$.
(v) By (ii) and (iv), Eq. (2.33) is confirmed.

2.2 Show (2.26). It is trivial that if $\boldsymbol{x} \in \mathrm{Ker}(B)$ then $\boldsymbol{x} \in \mathrm{Ker}({}^tBB)$.
Note that for $\boldsymbol{x} \in \mathrm{Ker}({}^tBB)$, it holds that $\langle \boldsymbol{x}, {}^tBB\boldsymbol{x}\rangle_{\mathbb{R}^n} = \langle B\boldsymbol{x}, B\boldsymbol{x}\rangle_{\mathbb{R}^m} = 0$.

2.3 (Complement of the proof of Proposition 2.8)

(a) This is also an simple exercise of linear algebra.

T. Takiguchi, *The Least Square Solution to an Overdetermined System*, SpringerBriefs in Mathematical Methods, https://doi.org/10.1007/978-981-92-0064-1

Let $B = \begin{pmatrix} b_{11} & b_{12} & \dots & b_{1n} \\ b_{21} & b_{22} & \cdots & b_{2n} \\ \vdots & \vdots & \ddots & \vdots \\ b_{n1} & b_{n2} & \cdots & b_{nn} \end{pmatrix}$ and $C = \begin{pmatrix} c_{11} & c_{12} & \dots & c_{1n} \\ c_{21} & c_{22} & \cdots & c_{2n} \\ \vdots & \vdots & \ddots & \vdots \\ c_{n1} & c_{n2} & \cdots & c_{nn} \end{pmatrix}$.

(i) Calculate the (ji) component of BC for $1 \le i, j \le n$.
(ii) Calculate the (ij) component of ${}^tC\,{}^tB$ for $1 \le i, j \le n$.
(iii) Confirm that both components calculated (i) and (ii) coincide.
(iv) Confirm that the answer to (iii) proves Proposition 2.8.

(b) For $\widetilde{A}$, it holds that

$$\begin{aligned} \widetilde{A} &= \widetilde{A}A\widetilde{A} = {}^t(\widetilde{A}A)\widetilde{A} = {}^t(\widetilde{A}AA^\dagger A)\widetilde{A} = {}^t((\widetilde{A}A)(A^\dagger A))\widetilde{A} \\ &= {}^t(A^\dagger A)\,{}^t(\widetilde{A}A)\widetilde{A} = A^\dagger A\widetilde{A}A\widetilde{A} = A^\dagger(A\widetilde{A}A)\widetilde{A} = A^\dagger A\widetilde{A}, \end{aligned} \tag{A.40}$$

where we have applied, (β) in Definition 2.4 for the first equation, (γ) for the second, (α) for the third, ${}^t(BC) = {}^tC\,{}^tB$ (for $B, C \in M_n(\mathbb{R}^n)$) (Problem 2.3 (a)) for the fifth, (γ) for the sixth, and (α) for the last.

(c) There are several other proofs, for example, to show

$$A^\dagger = \widetilde{A}AA^\dagger = \widetilde{A} \tag{A.41}$$

and so on. Try to give your original proof.

2.4 Check (a)–(d) in Definition 2.4 by substituting $A^\dagger = ({}^tAA)^{-1}\,{}^tA$.

Chapter 3

3.1

(a) Divide both sides of the first equation in (3.2) by h. Then by the second equation in (3.2), tending $h \to 0$ proves Proposition 3.1.

(b) It is sufficient to prove (i) ⇒ (ii). By Proposition 3.1, if f is totally differential at x_0 then the constant α in (3.2) is given by $\alpha = f'(x_0)$. Therefore, the representation

$$f(x_0 + h) - f(x_0) = f'(x_0)h + (f(x_0 + h) - f(x_0) + f'(x_0)h) \tag{A.42}$$

gives the first equation in (3.2), where in the second term in the right-hand side is $g(h)$, that is,

$$g(h) = f(x_0 + h) - f(x_0) + f'(x_0)h. \tag{A.43}$$

By virtue of (A.43), it is not difficult to show the second equation in (3.2), $\lim_{h\to 0}\frac{g(h)}{h}=0$.

(c) This is a typical problem of an ordinary differential equation of separation of variables type. By dividing both sides of (3.11), we have

$$\frac{u'(t)}{u(t)}=r(t) \tag{A.44}$$

since we can assume that $u(t)>0$, otherwise, if $u(t_0)=0$ for some $t_0\geq 0$, then $u(t)=0$ for $t\geq t_0$.

Take an indefinite integral of both sides of (A.44) with respect to t and we have

$$\int\frac{u'(t)}{u(t)}\,dt=\int r(t)\,dt \tag{A.45}$$

The left-hand side of (A.45) is represented by

$$\int\frac{u'(t)}{u(t)}\,dt=\int\frac{du(t)}{u(t)} \tag{A.46}$$

by differential equation $du(t)=u'(t)\,dt$. By virtue of (A.45) and (A.46), we have

$$\log u(t)=\int r(t)\,dt. \tag{A.47}$$

This is an answer.

We remark that if the initial value $u(0)=u_0$ is given, then (A.46) is more concretely given by

$$\int_0^t\frac{u'(t)}{u(t)}\,dt=\int_{u_0}^{u(t)}\frac{du(t)}{u(t)}=\int_0^t r(t)\,dt. \tag{A.48}$$

Therefore, the answer to the initial value problem, (3.11) with $u(0)=u_0$, is given by

$$u(t)=u_0e^{\int_0^t r(t)\,dt}. \tag{A.49}$$

3.2 Although this is a typical problem of an ordinary differential equation of separation of variables type, the author recommends the following solution very natural to the phenomenon; since the phenomenon occurs only in the interval $[a,b]$, it is natural to take the sum over $x\in[a,b]$ rather than applying the general method to solve the ordinary differential equations of separation of variables type.

By (3.13), we have

$$\frac{dI(x)}{I(x)} = -f(x)\,dx, \tag{A.50}$$

since we can assume that $I(x) > 0 \neq 0$ for any $x \in [a, b]$. Taking the sum of both sides in (A.50) over $x \in [a, b]$, we have

$$\int_a^b \frac{I'(x)\,dx}{I(x)} = -\int_a^b f(x)\,dx. \tag{A.51}$$

Note that $dI(x) = I'(x)\,dx$ in (A.51). Application of the change of the variables $x \to I$ in the left-hand side of (A.51) yields (3.14).

3.3 (Uniqueness of the Radon transform)
Let us define the Fourier transform $\widehat{f}(\xi)$, $\xi = (\xi_1, \cdots, \xi_n) \in \mathbb{R}^n$, of a function $f(x)$, $x = (x_1, \cdots, x_n) \in \mathbb{R}^n$ by

$$\widehat{f}(\xi) = \int_{\mathbb{R}^n} e^{-ix\cdot\xi} f(x)\,dx, \tag{A.52}$$

where $x \cdot \xi = x_1\xi_1 + \cdots + x_n\xi_n$ is the inner product.

(a) We denote by $l(\theta, s)$ for $\theta \in [0, 2\pi)$, $s \in \mathbb{R}$ the line $l(\theta, s) = \{(x, y) = (s\cos\theta - t\sin\theta, s\sin\theta + t\cos\theta) \mid t \in \mathbb{R}\}$. For a two-variable function $f(x, y)$ We note that for a fixed $\theta \in [0, 2\pi)$, $Rf(\theta, s) = Rf(l(\theta, s))$ is well-defined for almost all $s \in \mathbb{R}$ and $Rf(\theta, s) \in L^1(\mathbb{R})$ as a function of $s \in \mathbb{R}$.

$$\widehat{f}(0, \eta) = \int_{\mathbb{R}} e^{-iy\eta}\,dy \int_{\mathbb{R}} f(x, \eta)\,dx, \tag{A.53}$$

for $\eta \in \mathbb{R}$. Since $\int_{\mathbb{R}} f(x, y)\,dx = Rf(l(\frac{\pi}{2}, y)) = 0$ by assumption, we obtain that $\widehat{f}(0, \eta) = 0$. In the same way, we can prove for any $r \in \mathbb{R}$, $\widehat{f}(r\cos\theta, r\sin\theta) = 0$ by $Rf(\theta, s) = 0$ for any $s \in \mathbb{R}$. Therefore, $\widehat{f}(\xi, \eta) \equiv 0$ holds. The uniqueness of the Fourier transforms prove the uniqueness of the Radon transform.

(b) Let $f(x, y) = f_1(x, y) - f_2(x, y)$ and apply the solution to (a).

(c) We can generalize the solution to (a) for the general dimensional case. Let $\xi = (\xi', \xi_n) \in \mathbb{R}^n$ with $\xi_n \in \mathbb{R}$.

$$\widehat{f}(0, \xi_n) = \int_{\mathbb{R}} e^{-ix_n\xi_n}\,dx_n \int_{\mathbb{R}^{n-1}} f(x', x_n)\,dx' = 0, \tag{A.54}$$

where for $\xi \in \mathbb{R}^n$,

$$\widehat{f}(\xi) = \int_{\mathbb{R}} e^{-ix\cdot\xi} f(x)\, dx \tag{A.55}$$

is the Fourier transform of the function f. In the same way, we can prove that for any $\theta \in S^{n-1}$, the condition $Rf(H) = 0$ for any hyperplane $H \perp \theta$ implies $\widehat{f}(r\theta) = 0$ for any $r \in \mathbb{R}$. Since $\widehat{f}$ is continuous by Riemann–Lebesgue theorem, we conclude that $\widehat{f} \equiv 0$. Therefore $f \equiv 0$.

(d) Generalize the idea of the solution to (b).

3.4

(a) Review Sect. 3.1.3 well. Note that the line integral of the density distribution f along the line l_i is replaced by

$$a_{i1}x_1 + a_{i2}x_2 + \cdots + a_{in}x_n \tag{A.56}$$

by G. N. Hounsfield's idea.

(b) Note that the pixel function (3.18) is an approximation of $f(x, y)$ in $L^2(\mathbb{R}^2)$, that is,

$$\int_{\mathbb{R}^2} |f(x, y) - g(x, y)|^2\, dx\, dy \tag{A.57}$$

is small. On the other hand, since $\boldsymbol{x}^l$ in (3.23) is the least square solution to the system (3.21) (or (3.22)),

$$|x_j^l - x_j|^2 \tag{A.58}$$

are very small for all j's, which yields that $g(x, y)$ in (3.18) and $g_l(x, y)$ in (3.24) are very close in $L^2(\mathbb{R}^2)$, that is,

$$\int_{\mathbb{R}^2} |g(x, y) - g_l(x, y)|^2\, dx\, dy \tag{A.59}$$

is very small. Therefore, $f(x, y)$ and $g_l(x, y)$ are very close in $L^2(\mathbb{R}^2)$.

3.5 This problem is an example of stable type problems to be studied in Chap. 4.

(a) $\rho_0 = \sqrt{2}$.

(b) Note that the system (3.33) with $\rho = \sqrt{2}$ is the problem to find the origin as the point of the length $\sqrt{2}$ from the four points $(\pm 1, \pm 1)$.

(c) Subtracting both sides of the j-th equation from the counterpart in the i-th $(i < j)$ yields the semi-equivalent system of six linear equations.

(d) Apply Proposition 2.4 (iii), (2.19), to the system obtained in (c).

(e) Confirm both the solution to the system (3.33) with $\rho = \sqrt{2}$ and the least square solution obtained in (d) are the origin $(0, 0)$.

Chapter 4

As mentioned at the end of Chap. 4, the exercises in this chapter are difficult and include open problems and their solutions are omitted.

Chapter 5

5.1 Omitted.

5.2 Notice that the problem to give the least square solution to the system (5.14) is the one give the centroid of the triangle with the vertices $(-1, 0)$, $(1, 0)$ and $(0, 2)$. The answer to (e) is omitted.

Chapter 6

As mentioned at the end of Chap. 6, the exercises in this chapter are reviews of the context in the chapter, and their solutions are omitted.

Chapter 7

As mentioned at the end of Chap. 7, the exercises in this chapter are reviews of the context in the chapter, and their solutions are omitted.

Chapter 8

As mentioned at the end of Chap. 8, the exercises in this chapter are open problems and new research tasks, and their solutions are omitted.

Appendix A

A.1

(a) This proposition is proved by direct calculation. As others idea for the proof, note the following:

- If we multiply (ri) by $c \neq 0$, then multiplication by $\frac{1}{c}$ recovers the original matrix $(P_m(i;c)^{-1} = P_m(i;\frac{1}{c}))$.
- If we interchange (ri) and (rj) then the same operation recovers the original matrix $(Q_m(i,j)^{-1} = Q_m(i,j))$.
- If we add $c(rj)$ to (ri) then addition of $-c(rj)$ to (ri) recovers the original matrix $(R_m(i,j;c)^{-1} = R_m(i,j;-c))$.

(b) In order to transform the augmented coefficient matrix to the form (A.28) by the elementary operations (L1), (L2), (L3) and the interchange of two columns other than the last one, have the following order of the operations in mind:

(1) Make the $(1,1)$ component 1. Making use this, make the other components in the first column 0 by the operation (L3).

(2) Make the $(2,2)$ component 1. Making use of this, make the other components in the second column 0 by the operation (L3). In these operations, the first column, which was made the same as the unit matrix in the operation (1) must not be changed.

$\vdots$

(r) Make the (r,r) component 1. Making use this, make the other components in the r-th column 0 by the operation (L3). In these operations, the first, the second, $\cdots$, the $(r-1)$-th columns, which were made the same as the unit matrix in the operations (1), (2), $\cdots$, $(r-1)$ must not be changed.

The other part of the proof is given in the proof of Proposition A.2.

A.2 It is not difficult to show that $\boldsymbol{x} + \boldsymbol{y}$ and $\alpha\boldsymbol{x}$ are solutions to (A.37) for any two solutions $\boldsymbol{x}$, $\boldsymbol{y}$ to (A.37) and for any $\alpha \in \mathbb{R}$. To show that the dimension of the solution space is $n-r$, note that the set of the $n-r$ solutions

$$\begin{pmatrix} x_1 \\ x_2 \\ \vdots \\ x_r \\ x_{r+1} \\ x_{r+2} \\ \vdots \\ x_n \end{pmatrix} = \begin{pmatrix} -b_{1(r+1)} \\ -b_{2(r+1)} \\ \vdots \\ -b_{r(r+2)} \\ 1 \\ 0 \\ \vdots \\ 0 \end{pmatrix}, \begin{pmatrix} x_1 \\ x_2 \\ \vdots \\ x_r \\ x_{r+1} \\ x_{r+2} \\ \vdots \\ x_n \end{pmatrix} = \begin{pmatrix} -b_{1(r+2)} \\ -b_{2(r+2)} \\ \vdots \\ -b_{r(r+2)} \\ 0 \\ 1 \\ \vdots \\ 0 \end{pmatrix}, \cdots, \begin{pmatrix} x_1 \\ x_2 \\ \vdots \\ x_r \\ x_{r+1} \\ x_{r+2} \\ \vdots \\ x_n \end{pmatrix} = \begin{pmatrix} -b_{1n} \\ -b_{2n} \\ \vdots \\ -b_{rn} \\ 0 \\ 0 \\ \vdots \\ 1 \end{pmatrix} \tag{A.60}$$

is a base (or basis) of the linear subspace of the solutions to (A.37).

A.3 It is easy to prove {any solution to (A.7)} $\supset$ $\{\boldsymbol{x}_0 +$ (any solution to (A.31))$\}$. Here we prove {any solution to (A.7)} $\subset$ $\{\boldsymbol{x}_0 +$ (any solution to (A.31))$\}$

Take any solution $\boldsymbol{x}$ to (A.7), then it is represented as

$$\boldsymbol{x} = \boldsymbol{x}_0 + (\boldsymbol{x} - \boldsymbol{x}_0). \tag{A.61}$$

It is not difficult to show that $(\boldsymbol{x} - \boldsymbol{x}_0) \in$ {any solution to (A.31))}.

Therefore $\boldsymbol{x} \in \{\boldsymbol{x}_0 +$ (any solution to (A.31))}, which proves {any solution to (A.7)} $\subset \{\boldsymbol{x}_0 +$ (any solution to (A.31))}.

A.4

(a) If $\operatorname{rank}(A) = n$ then the following representation holds:

$$P_k \cdots P_1 A Q_1 \cdots Q_l = I_n, \tag{A.62}$$

where, $P_1, \cdots, P_k, Q_1, \cdots Q_l$ are elementary matrices. By Proposition A.1, elementary matrices are regular and their inverse matrices are also elementary matrices. Multiply $Q_l^{-1} \cdots Q_1^{-1}$ both sides of (A.62) from the right and then multiply $Q_1 \cdots Q_l$ from the left, which yields

$$Q_1 \cdots Q_l P_k \cdots P_1 A = I_n. \tag{A.63}$$

Equation (A.63) proves the proposition and the fact that the matrix A is regular with

$$A^{-1} = Q_1 \cdots Q_l P_k \cdots P_1. \tag{A.64}$$

(b) (i) $\Rightarrow$ (ii) is proved in the solution to (a). Let us prove (ii) $\Rightarrow$ (i) by contradiction. Assume that $\operatorname{rank}(A) = r < n$. In this case

$$P_k \cdots P_1 A Q_1 \cdots Q_l = F_{nn}(r), \tag{A.65}$$

where, $P_1, \cdots, P_k, Q_1, \cdots Q_l$ are elementary matrices. Since $P = P_k \cdots P_1$ and $Q = Q_1 \cdots Q_l$ are regular, we have

$$A = P^{-1} F_{nn}(r) Q^{-1}. \tag{A.66}$$

Since $F_{nn}(r)$ is not regular, the right-hand side of (A.66) is not regular, which contradicts the fact that A is regular.

References

1. Arakawa, M., Kanai, H., Ishikawa, K., Nagaoka, R., Kobayashi, K., Saijo, Y.: A method for the design of ultrasonic devices for scanning acoustic microscopy using impulsive signals. Ultrasonics **84**, 172–179 (2018)
2. Bliss, G.A.: Integrals of Lebesgue. Bull. AMS **24**, 1–47 (1917)
3. Censor, Y., Eggermont, P.P.B., Gordon, D.: Strong underrelaxation in Kacmarz's method for inconsistent systems. Numer. Math. **41**, 83–92 (1983)
4. Deans, S.R.: The Radon Transform and Some of Its Applications. Wiley, New York (1983)
5. Hashizume, K., Maruya, M., Ochi, T., Takabatake, T., Takiguchi, T.: Structure of the least square solutions to overdetermined systems and its applications to practical inverse problems. Japan J. Ind. Appl. Math. **41**, 945–960 (2024). https://doi.org/10.1007/s13160-023-00640-4
6. He, X.Q., Shu, Z.Q., Lid, Q.Y., Lu, G.Y.: Review of GPR rebar detection. In: PIERS Proceedings, Beijing, China, March 23–27, 2009, pp. 804–813 (2009)
7. Helgason, S.: The Radon Transform, 2nd edn. Birkhauser, Boston (1999)
8. Herman, G.T.: Image Reconstruction From Projections. Academic Press, New York (1980)
9. Hofmann-Wellenhof, B., Lichtenegger, H., Collins, J.: Global Positioning System, Theory and Practice, 5th edn. Springer-Verlag, Wiem, GmbH, New York (2001)
10. Maruya, M., Takiguchi, T.: Visual 3D reconstruction of a rotating object in space environment with a least-squares framework. In: Practical Inverse Problems and their Prospects, Mathematics for Industry, vol. 37, pp. 127–144. Springer Nature Singapore, Singapore (2023)
11. Mita, N., Takiguchi, T.: Principle of ultrasonic tomography for concrete structures and non-destructive inspection of concrete cover for reinforcement. Pac. J. Math. Ind. **10**(6) (2018). https://doi.org/10.1186/s40736-018-0040-0
12. Moore, E.H.: On the reciprocal of the general algebraic matrix. Bull. Am. Math. Soc. **26**, 394–395 (1920). https://doi.org/10.1090/S0002-9904-1920-03322-7
13. Natterer, F.: The Mathematics of Computerized Tomography, 2nd edn. Classics in Applied Mathematics, vol. 32. SIAM, Philadelphia (2001)
14. Ochi, T., Hashizume, K., Ishikawa, S., Takabatake, T., Takiguchi, T.: Recommendation to teach how to analyze an overdetermined system of linear equations with no solution in the class of linear algebra. In: Practical Inverse Problems and Their Prospects, Mathematics for Industry, vol. 37, pp. 111–126. Springer Nature Singapore, Singapore
15. Penrose, R.: A generalized inverse for matrices. Proc. Camb. Philos. Soc. **51**, 406–413 (1955). https://doi.org/10.1017/S0305004100030401
16. PS 300 FERROSCAN by HILTI, found at https://www.hilti.com/c/CLS_MEA_TOOL_INSERT_7127/CLS_CONCRETE_SCANNERS_7127/r9374764

T. Takiguchi, *The Least Square Solution to an Overdetermined System*, SpringerBriefs in Mathematical Methods, https://doi.org/10.1007/978-981-92-0064-1

17. Radon, J.: Über die Bestimung von Funktionen durich ihre Integralwarte längs gewisser Mannigfaltigkeiten. Ber. Verh. Sächs. Akad. Wiss. Leipzig, Math-Nat. **69**, 262–277 (1917)
18. Stein, E.M., Weiss, G.: Introduction to Fourier Analysis on Euclidean Spaces. Princeton University Press, Princeton (1971)
19. Takabatake, T., Hashizume, K., Ochi, T., Takiguchi, T.: Investigation of reinforcing bars in reinforced concrete structures by ultrasonic measurements. In: Practical Inverse Problems and Their Prospects, Mathematics for Industry, vol. 37, pp. 97–110. Springer Nature Singapore, Singapore (2023)
20. Takiguchi, T.: Ultrasonic tomographic technique and its applications. Appl. Sci. **9**, 1005 (2019). https://doi.org/10.3390/app9051005
21. Takiguchi, T.: Global uniqueness for the Radon transform. Bull. Korean Math. Soc. **57**, 597–605 (2020). https://doi.org/10.4134/BKMS.b190253
22. Takiguchi, T.: A theoretical study of the algorithm to practicalize CT by G. N. Hounsfield and its applications. Japan J. Ind. Appl. Math. **37**, 115–130 (2020). https://doi.org/10.1007/s13160-019-00391-1
23. Tanabe, K.: Projection method for solving a singular system of linear equations and its applications. Numer. Math. **17**, 203–214 (1971)
24. Tanaka, S., Wakabayashi, M.: On measurement of the depth and the diameter of steel bars in reinforced concrete using electromagnetic wave (Radar). In: 2006 SICE-ICASE International Joint Conference, pp. 2555–2559 (2006)
25. Weir, A.J.: Lebesgue Integration and Measure. Cambridge University Press, Cambridge (1973)
26. Xu, G.: GPS — Theory, Algorithms and Applications, 2nd edn. Springer, Berlin Heidelberg (2007)
27. Zalcman, L.: Uniqueness and nonuniqueness for the Radon transform. Bull. Lond. Math. Soc. **14**, 241–245 (1982)

Index

T. Takiguchi, *The Least Square Solution to an Overdetermined System*, SpringerBriefs in Mathematical Methods, https://doi.org/10.1007/978-981-92-0064-1

FSC
www.fsc.org
MIX
Papier | Fördert
gute Waldnutzung
FSC® C083411

Kolibri 360 GmbH
Ferdinand-Jühlke-Straße 7
99095 Erfurt, Deutschland
produktsicherheit@kolibri360.de